Women in Engineering and Science

The Springer Women in Engineering and Science series highlights women's accomplishments in these critical fields. The foundational volume in the series provides a broad overview of women's multi-faceted contributions to engineering over the last century. Each subsequent volume is dedicated to illuminating women's research and achievements in key, targeted areas of contemporary engineering and science endeavors. The goal for the series is to raise awareness of the pivotal work women are undertaking in areas of keen importance to our global community.

Caroline D'Amours • Hannah Young
Catherine Mavriplis

The Chairs for Women in Science and Engineering Program

Thirty Years of Action Across Canada, from 1989 to 2021

Springer

Caroline D'Amours
Department of History
Royal Military College of Canada
Kingston, ON, Canada

Hannah Young
University of Ottawa
Ottawa, ON, Canada

Catherine Mavriplis
Department of Mechanical Engineering
University of Ottawa
Ottawa, ON, Canada

ISSN 2509-6427 ISSN 2509-6435 (electronic)
Women in Engineering and Science
ISBN 978-3-031-54084-4 ISBN 978-3-031-54085-1 (eBook)
https://doi.org/10.1007/978-3-031-54085-1

This Springer imprint is published by the registered company Springer Nature Switzerland AG
The registered company address is: Gewerbestrasse 11, 6330 Cham, Switzerland

Paper in this product is recyclable.

In memory of
F. Mary Williams
and
Margaret-Ann Armour

Foreword

When I went to school I learned to write and how to read,
Some history, geography and home economy. ...
Then they had the nerve to say, 'What would you like to be?'
I says, 'I'm gonna be an engineer!'
No, you only need to learn to be a lady,
The duty isn't yours for to try and run the world.
An engineer could never have a baby!
Remember, dear, that you're a girl

– *From* I'm Gonna Be An Engineer *by Peggy Seeger*

The year 1989 was a pivotal one for women in science and engineering in Canada. And I use the word "pivotal" both literally and figuratively. It was the year when everything changed.

Before 1989, Canada had known more than a century of scientific advancement. Though a young country, Canada quickly established a reputation for itself, first as an importer and adopter of new technologies from abroad, and then as an innovator and developer of new scientific and technological capabilities in its own right. Women were important contributors to this reputation as both pure researchers and applied practitioners of science. But those women often received little support for their research or credit for their contributions. The history of science in Canada, while magnificent in many ways, is also a story of gender discrimination, replete with stories just like the one quoted above in Peggy Seeger's song: female scientists whose careers were stifled by convention and hampered by institutional sexism.

Consider the story of Harriet Brooks, who was the first woman to receive a Master's degree from McGill University in 1901. As the first graduate student under the tutelage of the vanguard nuclear physicist Ernest Rutherford, Brooks made numerous discoveries regarding radioactivity and the structure of the atom. In 1906, while teaching physics at Columbia University's Barnard College, she became engaged to a colleague – only to be told by Barnard's dean that "whenever your marriage does take place it ought to end your official relationship with the college." Brooks broke off the engagement to continue her career. But despite continued collaborations with Rutherford, as well as with Marie Curie in Paris, within a year

Brooks had concluded that she would likely be unable to find long-term employment in physics research, and abandoned her scientific work to become a wife and mother.

Or consider the experience of geologist Alice Wilson, whose employer of 54 years, the Geological Survey of Canada, repeatedly refused to grant her a leave of absence to pursue her doctoral studies – even upon learning that she had won a scholarship from the Canadian Federation of University Women. Her request was finally granted and, in 1929, she earned her PhD from the University of Chicago. Her requests to conduct field work were also denied on the grounds that women were barred from traveling to remote locations with men. Wilson overcame that obstacle by conducting solo fieldwork in the sparsely studied Ottawa–St. Lawrence Valley on foot and by bicycle. The Geological Survey routinely purchased automobiles for her male colleagues but refused one for her. Undaunted, she bought her own.

Dr. Wilson's story ends, thankfully, on a note of recognition. In 1938 she became the first female Fellow of the Royal Society of Canada, and in 1960 became the first woman to receive an Honorary Doctorate from Carleton University. But for every Alice Wilson who persevered there were many more Harriet Brookses, promising women whose talent and ambition stalled in the face of systemic discrimination. Throughout the 1970s and 1980s, the women's liberation and feminism movements opened more doors for women to pursue degrees and careers in science, while also drawing attention to the barriers that hindered women from advancing in the profession and reaching their full potential.

The summer of 1989 marked an important breakthrough for women scientists in Canada. That is when the Natural Sciences and Engineering Research Council announced the creation the Chair for Women in Engineering – the program that is the subject of this book. The position was created to address gender imbalance, with its stated goal being to increase the participation of women in engineering and to provide role models for women active in, and considering careers in, these fields. Biomedical engineer Dr. Monique Frize was appointed as the inaugural Chair at the University of New Brunswick.

Then the final month of 1989 marked a gruesome tragedy: the December 6 Montreal Massacre, in which a lone gunman, motivated by a deep misogyny, entered the École Polytechnique and shot and killed 14 female students while wounding 13 others, ten women and three men. It's no overstatement to say that the grief of December 6 is still very present. But in due course, that grief has also given rise to a broad recognition that women were under-represented, inadequately supported and discriminated against in the sciences and engineering – and to a determination to rebalance the scales.

For her part, Dr. Frize understood immediately how the events of December 6 would impact her work as an NSERC Chair. "My first day in the position was December 11, 1989, the day of the funeral for several of the victims," she has recounted. "I went to the funeral in Montreal and returned to Fredericton to give a talk to around 400 people at University of New Brunswick. What I said very determinedly was that we must have 1,000 more women engineers for every one who had been killed."

Since then, there has been a concerted effort to apply an evidence-based lens to the issue of women in the sciences, by collecting more data and tracking measurable progress. Between 1995 and 2022, the percentage of NSERC Discovery grants awarded to women more than doubled, from 9.3 percent to 25.1 percent, while the percentage of NSERC postdoctoral fellowships awarded to women rose from 17.3 percent to 40.4 percent.

Similar increases have shown up in labour force participation data, where the number of women in natural sciences and engineering occupations in Canada increased from 19.2 percent in 1995 to 24.4 percent in 2020.

And yet, even as gains are made, they are often lost in the course of women scientists' career progressions. Although women outnumber men in undergraduate university enrollments in Canada, fewer women than men enroll in STEM programs, and at every subsequent stage their proportion gets smaller still: graduate enrollment, faculty positions, middle and senior management in both the public and private sectors – a leaky pipeline with multiple holes throughout its length.

In 2010 the Association of Professional Engineers and Geoscientists of Alberta set a goal to raise the percentage of newly licensed engineers who are women to 30 percent by 2030. The goal, aptly named 30 by 30, was soon adopted nationally by Engineers Canada. And progress against that goal has been slow and steady: in 2014 women made up 17 percent of newly minted engineers, and by 2020 it had risen to over 20 percent. The 30 percent figure is widely acknowledged as a tipping point, driving the cultural shift necessary for sustainable, structural change at all career stages.

But while data can provide trend lines and measurable goals, it cannot provide inspiration. That can only come from women themselves: scientists who are recognized as experts in their fields, whose talents are acknowledged by their institutions, whose research is generously funded and whose leadership is allowed to blossom. These are the role models young women need when they are considering a career in the sciences; proof positive that their talent and hard work will be rewarded.

And therein lies the importance of the NSERC Chairs for Women in Science and Engineering. The program was expanded in 1996 from a single national Chair to a regional structure, with one Chair each in the Atlantic region, Quebec, Ontario, the Prairies and British Columbia. It also broadened its scope to include Chairs in the sciences as well as engineering. The program, which has enjoyed strong financial backing from industry throughout its history, is now welcoming its fourth generation of Chairs. Meanwhile, its alumnae have gone on to other leadership positions as heads of research organizations, engineering deans and university presidents.

This book tells their stories: their scientific discipline, their unique perspectives and their determination to make a difference – both individually and together. And through them this book tells the story of the NSERC Chairs for Women in Science and Engineering Program and the multitude of projects and initiatives its Chairs have undertaken to help women gain their rightful place in the scientific community.

When the next generation of young women seek out information on the history of women in Canadian science, they won't just discover the trials of Harriet Brooks and Alice Wilson. They'll also discover the triumphs of Monique Frize and her colleagues. They won't merely imagine that a better world might be possible for women in science. They'll also find tangible evidence that it's already taking shape, and know that they can be a part of it.

Chief Science Advisor of Canada
Government of Canada
Ottawa, Canada

Mona Nemer

Acknowledgements

The work to put together this thirty-year retrospective was a group effort. Arianne Zamin and Hannah Young did the first historical research, leading to Hannah starting the writing and Caroline D'Amours writing the bulk of the book. Catherine Mavriplis managed the overall project, wrote the final chapter and served as editor. However, each of the Chairholders gave generously of their time by giving interviews, providing reports, editing the sections about their Chair programs and fact-checking. The research was greatly facilitated by the Canadian Archive of Women in STEM, conveniently located at University of Ottawa. Many thanks to the excellent staff at the archive and some last minute fact-checking by Marie Noël. We acknowledge the tremendous support of the Natural Sciences and Engineering Research Council of Canada for its conception and excellent management of the Chairs for Women in Science and Engineering Program. In particular, we thank all the program managers, program officers and executives who supported the program over the years and were always attentive to the Chairholders' concerns and experiences. We also acknowledge all the sponsors of the individual Chairs: without their support the program would undoubtedly have been far less effective and influential. We also thank all assistants, participants and stakeholders who enthusiastically contributed to the Chair program. Support from Canada's Ministers of Science and other policymakers is gratefully acknowledged. Lastly, we thank Mona Nemer, Canada's Chief Science Advisor, for taking the time to write the foreword and for supporting the Chair program and participating in many Chair activities over the years. Many thanks to the professional team of people at Springer for accepting our proposal and making the publication process smooth.

Contents

Abbreviations

ACEC	Association of Consulting Engineers of Canada
ACFAS	Association francophone pour le savoir
ACM	Association for Computing Machinery
ACM-W	Association for Computing Machinery – Women
AFFESTIM	Association de la francophonie à propos des femmes en sciences, technologies, ingénierie et mathématiques
ANSWERS	Association of Nova Scotia Women for Education and Research in Science
APEC	Asia-Pacific Economic Cooperation
APEGA	Association of Professional Engineers and Geoscientists of Alberta
ARC	Alternate Routes to Computing
ASEAN	Association of Southeast Asian Nations
AUCC	Association of Universities and Colleges of Canada
BC	British Columbia
CAE	Canadian Academy of Engineering
CARIS	Collaborative Advanced Robotics and Intelligent Systems Laboratory
CBC	Canadian Broadcasting Corporation
CCPE	Canadian Council of Professional Engineers
CCWE	Canadian Committee on Women in Engineering
CCWEST	Canadian Coalition for Women in Science, Engineering and Technology
CCWESTT	Canadian Coalition for Women in Science, Engineering, Technology and the Trades
CEGEP	Collège d'enseignement general et professionnel
CERC	Canada Excellence Research Chairs
CIHR	Canadian Institutes of Health Research
CIPWIE	Committee for Increasing the Participation of Women in Engineering
CIRST	Centre interuniversitaire de recherche sur la science et la technologie

CISE	Chairs for Inclusion in Science and Engineering
CIWES	Canadian Institute of Women in Engineering and Science
CMA	Canadian Manufacturers' Association
CODE	Council of Ontario Deans of Engineering
COM-B	Capabilities, Opportunities, Motivation – Behaviour change theory of change
COVID	Coronavirus Disease
CRA	Computing Research Association
CRC	Canada Research Chairs
CREATE	Collaborative Research and Training Experience
CRSNG	Conseil de recherche en sciences naturelles et en génie
CSE	Computer Science and Engineering
CSFG	Chaire pour les femmes en sciences et en génie
CSME	Canadian Society for Mechanical Engineering
CURA	Community University Research Alliance
CWSE	Chair for Women in Science and Engineering
CWSE-ON	Chair for Women in Science and Engineering for Ontario
CWSEA	Chair for Women in Science and Engineering for the Atlantic
DAWEG	Division for the Advancement of Women Engineers and Geoscientists
DLR	German Aerospace Centre
DMP	Distributed Mentor Project
DNA	Deoxyribonucleic acid
EDI	Equity, Diversity and Inclusion
EES	Engendering Engineering Success
ESS	Engendering Success in STEM
FAFS	Faculty of Agricultural and Food Sciences at University of Manitoba
FFDC	The Bruce D. Campbell Farm and Food Discovery Centre
FIRST (Lego)	For Inspiration and Recognition of Science and Technology
FORWARD	Focus on Reaching Women for Academics, Research and Development
FRQNT	Fonds de la recherche du Québec – Nature et technologies
GPS	Global Positioning System
HTML	HyperText Markup Language
IBM	International Business Machines
ICWES	International Conference of Women Engineers and Scientists
INRS	Institut national de la recherche scientifique
INWES	International Network of Women Engineers and Scientists
IT	Information Technology
LAMH	Laboratoire de Machines Hydrauliques
LGBTQ2+	Lesbian, Gay, Bisexual, Transgender, Queer and Two-Spirit Plus
NAC	National Action Committee on the Status of Women
NCDEAS	National Council of Deans of Engineering and Applied Science
NRC	National Research Council of Canada

NSERC	Natural Sciences and Engineering Research Council of Canada
NSF	National Science Foundation (US)
OECD	Organisation for Economic Cooperation and Development
OIQ	Ordre des ingénieurs du Québec
ONWiE	Ontario Network of Women in Engineering
OPUS	Outils pédagogiques utiles en science
OSPE	Ontario Society of Professional Engineers
PEG NL	Professional Engineers and Geoscientists Newfoundland and Labrador
PEI	Prince Edward Island
RDI	Research Development Initiative
RIM	Research In Motion
RNA	Ribonucleic acid
SET	Science, Engineering and Technology
SFRI	Steering and Funding of Research Institutions
SFU	Simon Fraser University
SSHRC	Social Sciences and Humanities Research Council of Canada
STEM	Science, Technology, Engineering and Mathematics
SWIFT	Supporting Women in InFormation Technology
UBC	University of British Columbia
UFA	University Faculty Awards
UNB	University of New Brunswick
UNESCO	United Nations Educational, Scientific and Cultural Organization
UQTR	Université du Québec à Trois-Rivières
WEPAN	Women in Engineering and Programs Advocates Network
WFA	Women Faculty Awards
WinSETT	Women in Science, Engineering, Technology and the Trades
WISE	Women in Science and Engineering
WISEAtlantic	Women in Science and Engineering Atlantic
WISEN	Women in Science and Engineering Network
WISEST	Women in Scholarship, Engineering, Science and Technology
WRDC	Women in Resource Development Committee
WWEST	Westcoast Women in Science, Engineering and Technology

Chapter 1
A National Chair for Women in Engineering (1989–1997)

On May 12, 1989, a press conference at the University of New Brunswick (UNB) announced the creation of the Northern Telecom/Natural Sciences and Engineering Research Council (NSERC) Chair for Women in Engineering. This announcement would launch an entirely new concept, a university-based Chair devoted strictly to the recruitment and promotion of women in this field. This new program was to have a tenure-track professorship on the theme of women in engineering, supported financially for a five-year term by an industrial partner. Its specific goal was to increase the participation and retention of women in engineering and to provide role models for women active in, and considering, careers in this field. At the press conference, Arthur W. May, president of NSERC from 1986 to 1990, Claudette MacKay-Lassonde, Director of Education Relations for Northern Telecom and the first female vice president of NSERC's Board of Directors, and Frank Wilson, Dean of Engineering at UNB, talked about the new position. The successful candidate would need to be a woman engineer with a graduate degree. The Chairholder was expected to contribute up to 50% of her time to the activities of the Chair and the remaining time to her professor/researcher activities at the university. Consequently, she would automatically become a role model for young women who were considering or were already enrolled in an engineering program of studies.[1]

The new Chair program was the result of combining forces evolving throughout the 1970s and the 1980s. In Canada, the women's movement, the women in engineering movement, and the political agenda of the day, together with the strong leadership of women pioneer activists, all interacted together to give birth to a groundbreaking program that would direct its efforts toward attracting and promoting women in engineering from that point on. This chapter brings to light these different forces. It also presents the work of the first national Chairholder, Monique

[1] Monique Frize, *A Woman in Engineering: Memoirs of a Trailblazer*, Ottawa, University of Ottawa Press, 2019, p. 111.

C. D'Amours et al., *The Chairs for Women in Science and Engineering Program*, Women in Engineering and Science, https://doi.org/10.1007/978-3-031-54085-1_1

Frize, from 1989 to 1997. At the very start of her mandate, Monique Frize was confronted by an excruciating national experience which would have long-lasting impact on her work as Chair. Despite this, she would build a strong program in addition to demonstrating the need to expand this innovative initiative.

Canadian Women in Science and Engineering (1970–1989)

Similarly to other countries in the modern world, Canada witnessed in the late 1960s and the 1970s a resurgence of feminist activity. Women's movements during this period aimed to empower women in both their private and public lives. In Canada, this "second-wave feminism" challenged assumptions about conventional roles for women, fought discrimination, and pursued various strategies to provide equal opportunities for women in the paid labour force. In response to a month-long campaign by a coalition of 32 women's groups led by the Canadian Federation of University Women, Prime Minister Lester B. Pearson established the Royal Commission on the Status of Women in 1967. The report tabled in December 1970 presented 167 recommendations for reducing gender inequality across the various spheres of Canadian society. To ensure the implementation of these recommendations, the National Action Committee on the Status of Women (NAC) was created in 1972.

Over the next two decades, the NAC became the main feminist pressure group in the country and the official voice of a multitude of member and affiliated groups supporting gender equality in the job market and fighting for the eradication of wage discrimination. Indeed, girls and women were still facing obstacles in job training and education and were most of the time paid less than a man for the same job. As a result, the NAC and other feminist groups lobbied for measures that would give women access to occupations traditionally held by men. This pressure led to the establishment of several advisory councils, status of women committees, and other women's organizations committed to equality politics. With the inclusion of the principles of gender equality and employment equity in the Canadian Charter of Human Rights and Freedoms of 1982, the federal government and several of its provincial counterparts adopted pay and employment equity laws and policies. From that point on, several activists within the women's movement fought successfully for the implementation of state-supported initiatives directed at raising the number of girls studying science and engineering.[2]

[2]Veronica Strong-Boag, "Women's Movements in Canada: 1960–85", in *The Canadian Encyclopedia*, September 12, 2016, accessed May 17, 2020, https://www.thecanadianencyclopedia.ca/en/article/womens-movements-in-canada-196085; Ruby Heap, "Fighting the 'Corset of Victorian Prejudice': Women's Activism in Canadian Engineering during the Pioneering Decades (1970s–1980s)", in Catherine Carstairs (ed.), *Feminist History in Canada: New Essays on Women, Gender, Work, and Nation*, Vancouver, UBC Press, 2013, 220 and Heap, "Writing Them Into History: Canadian Women in Science and Engineering since the 1980's", in Andrea Martinez and Meryn Stuart (eds.), *Out of the Ivory Tower: Feminist Research for Social Change*, Toronto, Schumach Press, 2003, 50–51.

By the latter half of the decade, combined with the growing influence of the second-wave feminist movements and the elimination of the more obvious legal and administrative barriers to equal opportunities for women in the paid labour force, the measures adopted by federal and provincial governments started to show some results. The proportion of Bachelor's degrees granted to women in engineering and the applied fields increased from 3% in 1975 to 13% in 1989. While the proportion of women obtaining graduate degrees remained small during the same time period, the number of women earning Master's and doctoral degrees rose from 37 to 196 and from 9 to 19, respectively.[3] As a result, increasing numbers of women entered the job market: the number of women registered as professional engineers in Canada went from 463 in 1980 to 3,464 in 1989. However, this increase did not drastically change the composition of the profession. The proportion of women registered as professional engineers in the country rose from just 0.6% in 1980 to 2.9% 9 years later. So, while the 1970s and 1980s certainly saw an increase in women studying and working in engineering, as well as more women in general working in male-dominated fields, women were still clearly the minority[4] (Table 1.1).

At the same time, Canadian women in science and engineering began to voice their discontent about the barriers preventing them from fully accessing and participating in their respective fields and started fighting for equal rights in their professional life. As historian Ruby Heap noted, a growing number of them "were starting to question some of the practices and policies developed by the homogenous male engineering population."[5] In order to address these concerns, and to offset women's

Table 1.1 Women registered as professional engineers in Canada, 1980–1990/year end

Year	Total engineers	Women engineers	% Women
1990	121,464	3875	3.2
1989	118,335	3464	2.9
1988	115,562	3098	2.7
1987	113,032	2687	2.4
1986	109,898	1670	1.5
1985	107,205	2049	1.9
1984	105,703	1144	1.1
1983	103,109	956	0.9
1982	100,227	672	0.7
1981	96,749	567	0.6
1980	87,728	463	0.5

Source: Canadian Engineering Human Resources Inventory

[3] Heap, "Fighting the 'Corset of Victorian Prejudice'", 221–222.

[4] The Canadian Committee on Women in Engineering, "Table 2" in, "More Than Just Numbers: Report of the Canadian Committee on Women in Engineering," (Fredericton, N.B.: Northern Telecom-NSERC Women in Engineering Chair, Faculty of Engineering, University of New Brunswick, 1992).

[5] Heap, "Fighting the 'Corset of Victorian Prejudice'", p. 222.

exclusion from the male-dominated associations and organizations, women started building their own institutional framework. The alternative support network movement in science and engineering started in 1977 with the creation of Women in Science and Engineering/Femmes en Science et en Ingénierie (WISE). Founded by Claudette MacKay-Lassonde, a top-level manager at Hydro-Ontario at the time, the organization brought together women scientists and engineers "to encourage women to seek higher education and achievements, and to assist women entering into science or engineering."[6] This was done mostly by giving mutual support for women practitioners and by providing role models and mentorship to female students in science and engineering. Over the following decade, WISE had attracted more than 400 members, and local chapters were formed in Ontario, Quebec, Newfoundland, and New Brunswick.[7]

Claudette's pioneering role in promoting the entry and advancement of more women in the engineering profession did not stop there. She spearheaded the first Canadian Convention of Women Engineers in May 1981. Professional engineers, recent graduates from engineering schools, and engineering students from every province joined to discuss day care, stress on the job, and male animosity. The final formal recommendation of the convention was to make WISE a national organization with the mission to bring more women into engineering. Thanks to Claudette's efforts to break the isolation of women engineers, several advocacy committees, support networks, women's caucuses in universities and professional associations were established throughout the 1980s. These alternative organizations were engaged in various strategies designed to increase the participation of girls and women in science and technology, to promote equal opportunities for women scientists and engineers, and to counteract gender stereotyping. The movement gained momentum with the help of Claudette, the then president of Professional Engineers of Ontario, and Micheline Bouchard, a rising executive at Hydro-Québec and the president of the Ordre des Ingénieurs du Québec. It had a major impact on the engineering profession. Their call was heard by the Canadian Council of Professional Engineers (CCPE), the national body of the provincial engineering associations, and the National Council of Deans of Engineering and Applied Sciences: these two organizations worked to bring women engineers together and attract more women to the profession.[8]

---

[6] Correspondence with Valerie Davidson, by Hannah Elizabeth Young, email, May 7, 2018.

[7] Heap, "Fighting the 'Corset of Victorian Prejudice'", p. 222–223 and Heap, "Writing Them Into History", 51.

[8] Heap, "Fighting the 'Corset of Victorian Prejudice'", p. 223–225.

The receptiveness to addressing the problem of women in engineering arose also from the political context of the 1970s and 1980s. As discussed earlier, the federal government introduced state-supported initiatives aimed at women during these two decades. At the same time, the Canadian government was becoming concerned about the impact of rapid technological change and the skill level of the country's workforce. Government officials were particularly troubled by the serious shortage of skilled workers and university-trained scientists and engineers predicted by the Ministry of Employment and Immigration as well as the Ministry of State for Science and Technology in the early 1980s. These concerns were shared by NSERC. Since the 1970s, NSERC had feared critical shortages of scientific expertise amid aging and retiring university faculty. With the announcement of a new National Science and Technology Policy in 1987, the federal government implemented a national system of innovation that would rely heavily on the use of science and technology research. Thus, Canada would be even more dependent on the training and retraining of a highly qualified workforce. This new initiative would finance several support programs designed to attract more young women to science and engineering. In short, throughout the 1980s, women were seen by the federal government as the solution to the "manpower" needs of the country.[9]

As historian Ruby Heap aptly noted, the women in engineering movement was able to take advantage of the Canadian government's growing concern with technological innovation and the creation of a skilled labor force. In 1985, Claudette MacKay-Lassonde was nominated to NSERC's Board of Directors before being selected as its first female vice president two years later. She also became a member of the newly created National Advisory Board on Science and Technology in 1987. Her association with these federal agencies, as well as her new position as the Director of Education Relations for Northern Telecom, a leading global supplier

[9] Heap, "Fighting the 'Corset of Victorian Prejudice'", p. 228–229.

of telecommunications systems, would help move the women in engineering movement forward. As a matter of fact, Claudette was closely involved with NSERC's most innovative program to support women engineers.[10]

The First National Chair and the Polytechnique Massacre (1989)

In January 1989, the Faculty of Engineering at UNB submitted an innovative (and unsolicited) proposal to NSERC for the creation of an NSERC Chair for Women in Engineering.[11] While UNB did suggest some activities that the Chair could engage in, including conducting research on women in engineering and holding workshops for teachers and guidance counselors, they intended the incumbent to largely decide what their Chair program would look like. UNB explained in their application that "the exact nature of the duties of the Chair in Women in Engineering relative to the furthering of an interest in engineering by women would vary according to the incumbent's strengths and capabilities."[12] UNB's application made it clear, however, that the Chair would strive to make engineering an attractive career choice for women and girls, and that teaching, research, the recruitment of students and development of programs would all be necessary and expected.[13] UNB explained that as of January 1989—when their application was submitted—the Faculty of Engineering had already begun "the process of identifying qualified persons to apply for the position."[14]

While UNB was working with NSERC to bring about the Chair, Monique Frize was in the process of completing her doctorate in Clinical Engineering at the Erasmus University Rotterdam in the Netherlands, while also working as a clinical engineer for seven hospitals in eastern New Brunswick and serving an associate research professor at Université de Moncton. Prior to her graduate studies, Monique had a near two-decade-long career in the biomedical health care delivery industry, managing the development of medical devices. She had spent eight of those years working in Montreal, Quebec, followed by ten years in Moncton, New Brunswick.[15]

---

[10] "Pioneer for Women", Canadian Consulting Engineer: Magazine for Professional Engineers in Construction, October 1, 2000, accessed June 6, 2020, https://www.canadianconsultingengineer.com/features/pioneer-for-women/ and Heap, "Fighting the 'Corset of Victorian Prejudice'", p. 229–230.

[11] Proposal to NSERC by University of New Brunswick for support for a Professorship in Women in Engineering—Prepared by the Faculty of Engineering, pg. 1, January 1989, 10–23, Box: NSERC-UNB 2, Monique Frize Fonds, University of Ottawa Archives and Special Collections, Ottawa, Ontario, Canada.

[12] Ibid, 2.

[13] Ibid.

[14] Ibid, 3.

[15] Ibid.

Monique earned her undergraduate degree in Electrical Engineering from the University of Ottawa in 1966, where she was not only the solitary female in her class but also the first female to study engineering at the University of Ottawa.[16] As an undergraduate student, Monique remembers trying to draw as little attention as possible to herself, often dressing in very plain clothes in an attempt to fit in.[17] Monique describes her professors as "puzzled" by her, and they wondered "how do we treat this person?"[18] Eventually, they began to treat her like everyone else. Despite the initial bewilderment of her classmates and professors at having a woman in their classes, she remembers having a positive and supportive relationship with many of her classmates, although there was one particularly "nasty" student in electrical engineering.[19] Due to these experiences as an undergraduate, as well as her work in the health care industry, Monique was intimately aware of the challenges of learning and working in an almost entirely male-dominated field. Throughout her career, she made it a priority to encourage and support other women who were interested in taking the same path.

When UNB announced that they had been awarded the joint NSERC-Northern Telecom National Chair for Women in Engineering in May 1989 at a press conference, Monique was invited. The Dean of the Faculty of Engineering encouraged her to apply, as he had previously met Monique through her position at Université de Moncton. Monique remembers hearing the description of the candidate UNB was looking for, and thinking, "this sounds like superwoman." She applied anyway, since she figured "if they can't find a superwoman then I'll get the job."[20] When asked about what motivated her to apply, Monique explained that she felt ready for a new challenge, and academia provided the opportunity for her to bring her "real-life experience with medical devices into the classroom."[21]

[16] Monique Frize, interview by Hannah Elizabeth Young, telephone interview, May 7, 2018.

[17] Ibid.

[18] Ibid.

[19] Ibid.

[20] Ibid.

[21] Ibid.

On the condition of having completed her doctorate in August 1989, Monique was selected as the first Chairholder for a term of five years.[22] Although Monique's own experiences in engineering learning and working environments fueled the way she approached her new mandate, her program was also influenced by one particular perspective. After being appointed in November 1989, Monique called Professor Ursula Franklin at the University of Toronto to discuss her new position. A world-renowned physicist who pioneered the development of archaeometry, Ursula also had significant experience in researching the topic of women in engineering. While Ursula had never heard of Monique before this call, she was generous enough to give her opinion on the matter. The core advice Ursula gave to Monique was that the approach needed to be more than just numbers. As we have seen, Canadian engineering organizations expected labour shortages in the following years and were looking to attract women to the field to compensate for the low numbers. If she agreed that more women were needed in engineering, Ursula told Monique that women needed to be recruited first and foremost for what they could bring to the profession. Put differently, Ursula wanted the Chair to be about making engineering fit for women, rather than women fit for engineering.[23] Monique acknowledged later that "this thought was to inspire my entire work on women's issues for the next twenty years."[24]

Monique's program was also deeply influenced by a tragic event. A week before the start of her academic position at the University of New Brunswick, the École Polytechnique massacre, also known as the Montreal massacre, occurred, propelling the topic of women in engineering, and their rights and safety, to the forefront of Canadian media. On December 6, 1989, Marc Lépine entered the buildings of

[22] Ibid.

[23] Monique Frize, "Dr. Ursula Martius Franklin (1921–2016)", *Canadian Journal of the History of Science, Technology and Medicine*, vol. 41, no. 1 (2019), p. 68–69 and Frize, *A Woman in Engineering: Memoirs of a Trailblazer*, Ottawa, University of Ottawa Press, 2019, p. 113–114.

[24] Monique Frize, "Dr. Ursula Martius Franklin (1921–2016)", *Canadian Journal of the History of Science, Technology and Medicine*, vol. 41, no. 1 (2019), p. 68.

École Polytechnique, a French-language engineering school in Montreal, and proceeded to rush through several classrooms, separating the male students from the female, and shooting the female students. He murdered 14 young women and injured 14 others—ten women and four men—before killing himself. When he walked into one classroom, he told Nathalie Provost, who asked him why he was killing women, "you're all a bunch of feminists. I hate feminists."[25] Lépine proceeded to shoot Nathalie Provost, but she fortunately survived. The blatant hatred that Lépine showed toward women pursuing engineering careers, and women in general, shocked the nation and led to vigils being held in every major city in Canada. December 6 was named an official day of commemoration for female victims of male violence. Just days after the shooting, and still in her hospital bed, Nathalie Provost spoke to the press and encouraged women to study engineering: "I ask every girl in Quebec, every girl in the world, that if they want to be an engineer, to keep this idea in their mind."[26]

The Polytechnique massacre changed the field of engineering in Canada, as both the industry and engineering schools looked inward at how they treated women, and how the environment could be improved. On December 11, 1989, her first day on the job as NSERC—Northern Telecom Chair for Women in Engineering, Monique Frize was in Montreal. Instead of going to her office in Fredericton, she attended the funeral for the Montreal massacre victims.[27] She remembers walking out of the

[25] Wendy H. K. Chun, "Unbearable Witness: Toward a Politics of Listening," Differences: A Journal of Feminist Cultural Studies 11 no. 1 (1999) pg. 112.

[26] Francine Pelletier, "Montreal Massacre Victim Employed as Engineer—CBC Archives," CBC News, December 6, 1994, accessed June 29, 2018.

[27] Monique Frize, interview by Hannah Elizabeth Young, telephone interview, May 7, 2018.

church with Claudette MacKay-Lassonde and adopting a new goal for her position as Chair: "One thousand more women engineers for each one who died."[28] When asked later about how the Polytechnique massacre impacted Canada, Monique recognized that it really seemed to focus the national attention on women in engineering. While violence against women was a topic that was being addressed, the massacre really made the public and the engineering community think about the issues that women faced in the field, and what to do about them.

If Monique was already convinced of the importance of the Chair by her own personal experience and the advice given by Ursula Franklin, she admitted later that, after the massacre, her new position "became a mission, as it was urgent to spread the message to girls and women not to fear choosing engineering as their study program."[29] This painful event also compelled her to very quickly increase her knowledge on issues and strategies to promote the participation of women in engineering. Indeed, in the three months that followed, Monique pored over research "about math and girls, and gender issues, and the social issues, engineering life issues for women."[30] The Montreal massacre would also drive part of her program for the years to follow.

The More than Just Numbers Report (1990–1992)

On February 22, 1990, the Canadian Committee for Women in Engineering (CCWE) was formed in response to the Polytechnique massacre, with Monique as the Chair.[31] Its premise was based on the following question: as there were no barriers of a physical or intellectual nature preventing women from being engineers, why did they represent such a minority in the population of registered professional engineers? Given this phenomenon, the CCWE sought to "uncover the social and cultural barriers responsible for the under-representation of women in engineering and to design strategies that will bring them as full participants into the profession."[32] In order to fulfill this mandate in a timely fashion—the CCWE set a goal to release a

[28] Frize, *A Woman in Engineering: Memoirs of a Trailblazer*, Ottawa, University of Ottawa Press, 2019, p. 116.

[29] Frize, *A Woman in Engineering: Memoirs of a Trailblazer*, Ottawa, University of Ottawa Press, 2019, p. 119.

[30] Monique Frize, interview by Hannah Elizabeth Young, telephone interview, May 7, 2018 and Frize, *A Woman in Engineering: Memoirs of a Trailblazer*, Ottawa, University of Ottawa Press, 2019, p. 116.

[31] Ibid.

[32] Canadian Committee on Women in Engineering, "More Than Just Numbers Summary: Report of the Canadian Committee on Women in Engineering," (Fredericton, N.B.: Northern Telecom-NSERC Women in Engineering Chair, Faculty of Engineering, University of New Brunswick, 1992), pg. 1.

comprehensive report in 1992—the CCWE developed several strategies to conduct its research.

The CCWE committee members formed a diverse group of engineers from academia as well as industry, students, and non-engineers. There were 19 members of the committee, of which Monique was Chair, Nora Hockin and Danielle Plouffe were ex-officio members, and Jeanne Inch was the sole staff member of the committee. Other committee members included Donald Laplante, Executive Director of the Canadian Council of Professional Engineers (CCPE); Ken Williams, the past president of CCPE; Eva Kushner, from the Association of Universities and Colleges Canada; Richard Gordon, from the Canadian Manufacturers' Association; and Valerie Davidson, an associate professor in engineering at the University of Guelph and a member of WISE who would later become the Chair for Women in Science and Engineering in Ontario.[33] While the CCWE was founded partly in response to the Polytechnique massacre, it was actually Nora Hockin, who at the time was the Director of University Research Policy at Industry, Science and Technology Canada, who came up with the idea. She approached the CCPE—now Engineers Canada—and persuaded them to be signatories along with other key stakeholders. Four organizations—the CCPE, the Association of Consulting Engineers of Canada (ACEC), the Canadian Manufacturers' Association (now Canadian Manufacturers & Exporters, known then as CMA), and the Association of Universities and Colleges of Canada (AUCC)—"combined forces to create a national project to investigate the participation and environment for women in engineering in Canada."[34] Once these four organizations were assembled, Nora approached Monique to be Chair of the committee. Since she was already the National Chair for Women in Engineering, and the goals of the CCWE were in line with the Chair mandate, it made sense for her to be at the helm.[35]

One major initiative of the CCWE was to hold six public forums across Canada in 1990–1991. These were the Ottawa Public Forum, September 12, 1990; the West Coast Forum, November 16, 1990; the Montreal Forum, January 29, 1991; the Prairie Region Forum, February 26, 1991; the Ontario Public Forum, March 6, 1991; and the Atlantic Region Forum, March 26, 1991. The goal of these forums was to "identify the barriers and bridges to the participation of women in engineering."[36] Attendees of each forum made presentations, and over 200 briefs were received in writing. Briefs and presentations were presented by teachers, both elementary and high school, engineering professors and deans, engineering employers, engineering association representatives, and students, both male and female. Additionally, over 40 briefs were mailed to the CCWE separately from both women

[33] Ibid 10.

[34] Canadian Committee on Women in Engineering: A Background Document, pg. 1, 1990, 10–23, Box: CCWE 1, Folder: A Chairperson's guide to Industrial Adjustment Service (IAS) Committee, Monique Frize Fonds, University of Ottawa Archives and Special Collections, Ottawa, Ontario, Canada.

[35] Monique Frize, interview by Hannah Elizabeth Young, telephone interview, May 7, 2018.

[36] Ibid.

engineers and women engineering students, and those were all taken into account in the CCWE's final report.[37]

The second strategy that the CCWE took to fulfill its mandate was the commission of a professional research study on seven faculties of engineering and six engineering workplaces to better understand what practices and policies were successful, in recruiting, retaining, and advancing the careers of women engineers. The CCWE stated that "the findings [of the study] were consistent with the statements made by the participants in the public forums."[38] The CCWE also held a one-day stakeholder meeting in Ottawa in September 1991, with over 20 stakeholders representing various groups including elementary and secondary educators, universities, professional associations, and employers. Throughout the day, "the participants helped identify reasonable objectives and timeframes and suggested ways to monitor implementation."[39]

In May 1991, the CCWE organized the national conference, *Women in Engineering: More Than Just Numbers*, in Fredericton, New Brunswick. The conference brought together all significant stakeholders in the employment and education of women in engineering such as senior management in industry, students, human resource managers, and professors. The conference had two main goals, the first of which had two parts: "To present the Committee's collected information to date, gathered from public forums, briefs, trends, [and] research findings" and "to present the Committee's draft recommendations on solutions and implementation plan." The second goal of the conference was "to obtain public commitment for action by key stakeholders."[40] Over 240 people attended the CCWE conference, and of these attendees, two thirds were women and one third men.[41] Through a series of workshops at the conference, the attendees were able to provide feedback on the committee's proposed recommendations. The feedback was then used to produce the *More Than Just Numbers* report, which was sent to all major signatories in September 1991.

In April 1992, the *More Than Just Numbers* report was released to the public, outlining 29 recommendations, six key factors to the successful implementation of these recommendations and a five-year timeline to achieve five main goals: (1) girls

---

[37] Ibid.

[38] Canadian Committee on Women in Engineering, "More Than Just Numbers Summary: Report of the Canadian Committee on Women in Engineering," (Fredericton, N.B.: Northern Telecom-NSERC Women in Engineering Chair, Faculty of Engineering, University of New Brunswick, 1992), pg. 1.

[39] Ibid.

[40] Canadian Committee on Women in Engineering, "Draft 26/11/90: Women in Engineering: More Than Just Numbers," pg. 1, 1990, 10–23, Box: CCWE 1, Folder: A Chairperson's guide to Industrial Adjustment service (IAS) Committee, Monique Frize Fonds, University of Ottawa Archives and Special Collections, Ottawa, Ontario, Canada.

[41] Canadian Committee on Women in Engineering, "More Than Just Numbers Summary: Report of the Canadian Committee on Women in Engineering," (Fredericton, N.B.: Northern Telecom-NSERC Women in Engineering Chair, Faculty of Engineering, University of New Brunswick, 1992), pg. 10.

and boys will pursue mathematics and science in equal numbers, especially at advanced levels throughout high school; (2) women will comprise 25–35% of first-year students, 20% of Master's students, 10% of doctoral students, and 5% of the professorate in faculties of engineering across Canada; (3) women will comprise at least 18% of graduates from undergraduate engineering programs; (4) more women engineers will be in senior management positions and on boards of directors of companies employing engineers; and (5) more women engineers will be elected members of council and appointed members of committees of associations of professional engineers.[42] The CCWE proposed a five-year timeline because of the "window of opportunity" that was open to Canadian society due to the impending retirement of baby boomers, many of whom would be leaving their positions as school teachers. Turnover creates an opportunity for change, as new educators can be trained with different perspectives, and these new attitudes were seen to "have a profound impact on the career choices of girls and women for decades to come."[43]

The CCWE report was split into five sections: Laying the Foundation, which addressed pre-university years for girls, starting with early childhood social influences; Education for and by Women Engineers, which discussed how postsecondary institutions can create engineering environments that are friendly to women; Designing Engineering Workplaces, which addressed how changes can be made in the corporate environment; Support by Association, which discussed how professional associations can support and promote women engineers; and Making it Happen, which suggested ways to implement the various recommendations made throughout the previous four sections of the report.[44]

The *More Than Just Numbers* report also included many personal testimonies of those who participated in the conference and forums, as well as those who submitted briefs separately. One female engineer at the Ontario forum spoke of the happiness she felt to be in a room with so many other women engineers: "When I saw the line-up for tonight's speakers, it was the largest group of female professional engineers that I had ever seen. [...] Usually, a woman engineer must face all of this alone. She has very few peers to compare with, particularly at work. She is at best a minority, and at worst an oddball."[45] Several women spoke of the lack of female role models they had while in secondary and postsecondary schooling, while others spoke of the positive camaraderie they felt with their male coworkers. One woman provided advice about how to deal with coworkers who would expect you to take on a domestic role at the office: "To avoid serving coffee in the field office, bring your own thermos. To avoid serving coffee in the corporate board office, tell everybody

[42] Canadian Committee on Women in Engineering, "The Goals," in Women in Engineering: More than Just Numbers, (Fredericton, N.B.: Northern Telecom-NSERC Women in Engineering Chair, Faculty of Engineering, University of New Brunswick, 1992).

[43] Ibid.

[44] Ibid, "The Report."

[45] Ibid.

you (a) don't drink coffee; (b) don't know how to make coffee; or (c) you're drinking tea."[46]

The CCWE forums and conference provided women engineers with an opportunity to share their experiences, whether positive and negative, in a way many of them had never had before. They got to meet supportive stakeholders who were eager and willing to listen and take part in the processes of change. At the same time, it was clear that many women engineers had had negative experiences at work, and at times felt isolated and alone in a profession of mainly men. The CCWE concluded the personal testimonies section in the *More Than Just Numbers* report with the following question: "Are we being fair to girls and young women in the school system and in universities when we say engineering is a viable career for women? We believe the answer is 'yes', but a qualified 'yes'."[47] There was no reason to believe that changes could not be made to the engineering profession to make it women-friendly but change itself was essential.[48]

In the foreword of the *More Than Just Numbers* report, Monique wrote of what a meaningful experience being a part of the CCWE was, both for herself and for many other female engineers. Throughout the public forums and conference, and the reading of briefs, Monique began to draw many parallels between her experience and that of the women taking part in the process. She "began to remember incidents of harassment and discrimination, the public insults and private apologies."[49] Through her own recovery from "amnesia," she was able to better understand the denial that so many female engineers have—both as students and professionals—as they begin their careers. Monique was also positively affected by many of the testimonies as women spoke "with determination, humour and vitality about their love of engineering."[50] Both men and women also outlined what steps they were already taking to improve the engineering environment for all engineers and provided thoughtful suggestions and feedback on how to encourage women to join engineering and to stay in engineering. Monique closed the foreword by saying "we must build the bridges that lead to an increased participation of women in engineering. Women bring their own perspective which, when blended with the masculine, will enrich the technological solutions of tomorrow."[51]

The *More Than Just Numbers* report certainly affected the Canadian engineering community, but the report itself received a great deal of attention from the wider public. The CCWE issued multiple press releases at different stages in the report creation process, and media outlets also reported on it. The hope was for the report to receive as much attention as possible, in order for it to have maximum impact. The *More Than Just Numbers* report was backed by the four organizations that

[46] Ibid.

[47] Ibid, "Personal Testimonies."

[48] Ibid.

[49] Ibid, "Foreword."

[50] Ibid.

[51] Ibid.

formed the CCWE—CCPE, CPEP, CMA, and AUCC—and it had many other signatories throughout the country.[52] By and large, the engineering community in Canada was on board with the *More Than Just Numbers* report. Many universities implemented the recommendations outlined in the report, leading to changes such as no longer tolerating misogynistic content in university newspapers, or orientation activities like the Lady Godiva parade.[53] The time following the *More Than Just Numbers* report was a time of positive anticipation, as change was on the horizon, and many people were ready for it. After the release of the *More Than Just Numbers* report, the CCWE was disbanded for the time being, and as the National Chair for Women in Engineering, Monique became the "natural contact person for the continuation of the development of the process of change on these issues."[54]

Developing a Chair Program (1989–1992)

When Monique began as the national Chairholder in 1989, there was very little to build on in terms of programming, and no real comparisons to be made with other positions in the country, or in the world, as the Chair was the first of its sort in Canada, and the globe. Apart from some suggested programming ideas, and the general expectations that the Chair would conduct research, teach, recruit students, and run programs and events, UNB really left the direction the Chair would take up to the incumbent. As Monique describes it, she was building the Chair program from "scratch."[55] In an interview with *The Globe and Mail* in 1990, Monique framed her role as Chair as a marketing role; "women think we wear a hard hat and work in mud building bridges, but that's only 5 percent of it. The other 95 percent work with computers, with a pencil and paper, and people."[56] For Monique, it was about showing women and girls that engineering is a job that helps people, as well as providing role models to change the way young girls, guidance counselors, teachers, parents, and society view the field of engineering and engineers.

As Monique began to develop her program, outlining what activities and outreach would take place over the following five years, she started to receive a great

---

[52] Canadian Committee on Women in Engineering, "A Background Document," pg. 1, 1990, 10–23, Box: CCWE 1, Folder: A Chairperson's guide to Industrial Adjustment service (IAS) Committee, Monique Frize Fonds, University of Ottawa Archives and Special Collections, Ottawa, Ontario, Canada.

[53] Certain student organizations staged an annual "Godiva Ride" in which a naked female rode a horse across campus.

[54] Monique Frize, "Opening Remarks," 1992, 10–23, Box: CCWE 6, Folder: MTJN & CCWEST Conference, Monique Frize Fonds, University of Ottawa Archives and Special Collections, Ottawa, Ontario, Canada.

[55] Ibid.

[56] Orlando French, "Engineer Clears Path for Other Women," 1990, The Globe and Mail, 10–23, News Clippings Folder 2 of 2, NSERC-UNB Box 8, Monique Frize Fonds, University of Ottawa Archives and Special Collections, Ottawa, Ontario, Canada.

deal of media attention. The nature of the NSERC—Northern Telecom Chair for Women in Engineering was such a novel concept, and following the Polytechnique massacre, the media was very eager to hear about the program. The media was not only interested in the Chair position and role, but the incumbent, and Monique was passionate and eager to speak on the subject of women in engineering. One article in *The Telegraph-Journal* from November 1990 described Monique's first year as Chair as one full of travel, where she visited over 36 schools in New Brunswick to "spread the word that women belong in engineering."[57] When asked why she travelled to so many elementary schools, Monique responded that "getting them early was the key to the future of women in engineering" because it was so important that young girls learn early that "no career is closed to them."[58] Monique explained that while almost all of the female students she spoke to at the elementary schools she visited in 1990 knew what a doctor does, very few had any idea of what an engineer does. After telling these girls some of the wide variety of things an engineer might do, Monique stressed that all the students needed to stay in the math and sciences in high school, or they would cut themselves off from a huge number of careers. The article concludes with Monique saying that until women reach critical mass in engineering programs, she will continue "her pitch to young girls."[59]

Another article published in the *Ottawa Citizen* in 1990 was interested in trying to understand what exactly Monique, as Chair for Women in Engineering, and as Chair of the CCWE, was trying to do. Monique explained that she was asking questions, in an attempt to better understand why women were not entering engineering. Was it the image of a "cigar-chomping man, sporting a yellow hard hat" that kept women away from pursuing engineering studies?[60] Or perhaps it was misogynist engineering students who intimidated women in their classes? While Monique did not yet have the answers to those questions, she had plenty of reasons why women should go into engineering, including the predicted labour shortage, that engineering was an intellectually challenging and rewarding job, and that women can bring something unique to the field. Women can make products and designs that are more user-friendly and practical, like sidewalk grates that high heels cannot get stuck in, or computers that are easier to use. Monique also addressed the sexism that many female engineers face at school, or in the workforce. She explained that "improving a women's self-image is the best-weapon against others' attitudes," and that she had established peer-counselling and networking at UNB, so that women in engineering can support one another.[61] Being a role model and helping to create more role

[57] Sharon Pond, "She Wants to Swell the Ranks of Women Engineers," November 24, 1990, The Telegraph Journal, 10–23, News Clippings Folder 1 of 2, NSERC-UNB Box 8, Monique Frize Fonds, University of Ottawa Archives and Special Collections, Ottawa, Ontario, Canada.

[58] Ibid.

[59] Ibid.

[60] Shelley Page, "Women in Engineering," October 7, 1990, The Ottawa Citizen, 10–23, News Clippings Folder 1 of 2, NSERC-UNB Box 8, Monique Frize Fonds, University of Ottawa Archives and Special Collections, Ottawa, Ontario, Canada.

[61] Ibid.

models was a goal for Monique, which is why she always accepted opportunities to speak to the media, in the hopes that another woman or girl will hear about a female engineer and realize that she can be one too.[62]

Since speaking engagements occurred so frequently, Monique developed a "menu" of speeches. Monique spoke to elementary schools, high schools, universities, "to the Ministry of Transportation in Ontario, the Royal Bank, the BC Hydro, all over the country," and each type of engagement required a different speech.[63] While she was in Ottawa, she re-organized a program of volunteers called *Pathmakers*, who would travel to different schools from K-12 and give speeches about engineering.[64] Developing a speech prototype for each type of event was a way to stay organized, and ensure that whether it was Monique speaking, or volunteers, that they could deliver information that was appropriate for the demographic present. One day, Monique could be encouraging elementary and high school girls to stay in math and science, and the next day she could be speaking to university students about graduate studies and encouraging them to stay in engineering, or she could be speaking to corporations about what values female engineers can bring to their companies.[65] Monique also created presentations to prepare female engineers for the workforce, so they might know what challenges to expect and how to cope with them. While Monique wanted women to be informed of the potential issues they might face regarding harassment, discrimination, and equity, she did not want to scare them away from an excellent career.[66] Monique was away from Fredericton every ten days, often giving speeches and attending meetings. If she happened to be in a city for a speech, she would reach out to other universities in that city and offer to speak, to maximize the reach of the Chair, and to minimize the costs. One of the elements Monique made sure of was the accuracy of all the information she provided in her "menu" of speeches. As the only Chair for Women in Engineering in the country, she felt she had to build a very solid and reputable program and that all the information that was provided was "well-documented, and well-researched."[67]

Since 1989, Monique had continued to send volunteers—and later, her two assistants, Jane McGinn-Giberson and Gisèle Levesque—to conduct a series of school presentations throughout the province. In 1993, they gave a total of 106 school presentations, speaking to approximately 5,400 students. Throughout 1993, Monique and her assistants gave 43 speeches and public lectures, all over the country.[68] Monique also gave 46 media interviews, and in contrast to 1992, when almost all of the interviews given were on the subject of the CCWE, the interviews were

[62] Monique Frize, interview by Hannah Elizabeth Young, telephone interview, May 7, 2018.

[63] Ibid.

[64] Ibid.

[65] Ibid.

[66] Ibid.

[67] Ibid.

[68] Monique Frize, "1993 Annual Report," pg. 2, 1993, 10–23, Box: NSERC UNB Chair 1 1, Folder: 1993 Annual Report, Monique Frize Fonds, University of Ottawa Archives and Special Collections, Ottawa, Ontario, Canada.

generally focused on the Chair. The media was interested in what activities the Chair program was running. Later in the year, the media interviews were largely focused upon Monique's renewal as Chair, for another five-year term.[69]

Understandably, Monique's first years as Chair were dedicated predominantly to public awareness activities. However, she also worked to develop outreach initiatives. The *Worlds UNBound Science and Engineering Day Camp* launched in 1993 at UNB, involving both a summer camp and outreach workshops. *Worlds UNBound* was based upon a similar science discovery camp that was started by students at Queen's University in 1987 that eventually spread across Canada and became part of the national Youth Engineering and Science (YES) program, which was based in Ottawa. The purpose of *Worlds UNBound* was to eliminate stereotypes through specific objectives, one of which was to "provide the children with positive female and male role models in the fields of engineering and science."[70] The workshops were intentionally held province-wide so that both rural and urban students would be reached. According to the event report, "visiting smaller communities seems to be very worthwhile since these communities seem to be constantly overlooked."[71] The *Worlds UNBound* program was designed to ensure the participation of an equal number of girls and boys. More than 120 workshops were presented in various classrooms throughout the province of New Brunswick during the months of May and June, while the actual summer camps were held in July and August for students in Grades 5–8. Over 227 students attended the summer camps. The program continued for the remainder of Monique's time as Chair at UNB, and in 1995, it reached over 4,500 students province-wide. This award-winning, bilingual, non-profit self-sustaining program administered by engineering and science students at UNB has grown over time and proved to be very successful. For example, throughout the 2010s, the initiative regularly reached close to 14,000 youth per year through camps, workshops, the Quest 4 Girls club, and other programs.[72]

Some other outreach initiatives the Chair program developed included a video *Engineering: Design Tomorrow's World*, as well as a newsletter *Women in Engineering: Recruitment and Retention*.[73] The video was aimed at elementary and high school students and was distributed and sold throughout the country, in both official languages, in the hope that it could be shown at schools to inform students about what an engineering career could look like, and to encourage girls to be engineers. The video could also reach rural and remote schools that would otherwise not

[69] Ibid.

[70] Monique Frize, "Appendix E: 1995 Worlds UNBound Final Report," pg.7, Box: NSERC UNB Chair 1 1, Folder: 1995 Annual Report, Monique Frize Fonds, University of Ottawa Archives and Special Collections, Ottawa, Ontario, Canada.

[71] Ibid, pg. 11.

[72] Frize, *A Woman in Engineering: Memoirs of a Trailblazer*, Ottawa, University of Ottawa Press, 2019, p. 163.

[73] Monique Frize, "1993 Annual Report," pg. 4, 1993, 10–23, Box: NSERC UNB Chair 1, Folder: 1993 Annual Report, Monique Frize Fonds, University of Ottawa Archives and Special Collections, Ottawa, Ontario, Canada.

be accessible to outreach opportunities.[74] The newsletter, which would prove to be a challenge in the following years to produce and circulate, was aimed at keeping the engineering community, in particular engineering faculty and schools, up to speed with the status of women and engineering.[75] In terms of initiatives at UNB, the Chair office provided one-on-one counselling to women students and women engineers from the university as well as beyond. UNB had also announced the development of a "biomedical engineering undergraduate option," which Monique was optimistic would interest many students, especially women.[76] Monique was also chairing the ad hoc committee created in the Faculty of Engineering at UNB to "assess what CCWE recommendations have been implemented at UNB, identify which ones have not, and what priorities should be for the latter; an action plan is to be developed for their implementation."[77] Representatives from all engineering disciplines were on the ad hoc committee in order to ensure that all disciplines within the engineering faculty were on the same page. Just as the *More Than Just Numbers* report demonstrated, the recruitment of women in engineering was certainly essential to increasing their representation, but so was retention, and implementing the CCWE recommendations was paramount to both elements. It seems that Monique was pretty successful in her quest to bring the issue of women in engineering to the public through media interviews and presentations in schools around the province. Indeed, the Fredericton campus of UNB registered an increase from 11% to 18% of first-year engineering students who were women, while the Saint-John campus noted that 20% of first-year engineering students were female.[78]

By 1993, it was already four years into Monique's five-year tenure, and it was time to decide whether UNB, and Monique, would apply for a renewal. Many initiatives in the Chair program took months of planning to begin; such as *Worlds UNBound* and the *Engineering: Design Tomorrow's World*, and the ad hoc committee at UNB had just been formed. In March 1993, the university submitted a renewal application to NSERC for the Chair. The application stated that so far, the Chair had been a success: enrollment of women in engineering programs across Canada had increased, and in 1994, the representation of women reached 18.2%.[79] The numbers were certainly moving, and relatively rapidly, considering women represented just

[74] Ibid.

[75] Ibid.

[76] Ibid, pg. 6.

[77] Ibid, pg. 5.

[78] Frize, *A Woman in Engineering: Memoirs of a Trailblazer*, Ottawa, University of Ottawa Press, 2019, p. 150.

[79] NSERC Task Force on Women in Science and Engineering, "Towards Building a New Scientific and Engineering Culture in Canada," pg. 2, 1996, 10–23, Box: NSERC UNB Chair 1, Folder: Report of the Task Force on Women in Science and Engineering to the Natural Sciences and Engineering Research Council, Monique Frize Fonds, University of Ottawa Archives and Special Collections, Ottawa, Ontario, Canada; and "Percentage of Female Undergraduates in Engineering (Canada)," Ontario Network of Women in Engineering (ONWiE), accessed June 20, 2018, http://www.onwie.ca/resources-tools/statistics/percentage-of-female-undergraduates-in-engineering-canada-

13% of the engineering undergraduate population four years before. Monique had also become a role model for women and girls across the country, which was in itself a great success. Her achievements and efforts were noticed throughout Canada, and in 1993, Monique was made an Officer of the Order of Canada for her contributions to, and achievements in, the engineering community.[80] At the same time, as demonstrated in the CCWE report, much was left to be done. UNB requested a second five-year term in "order to build on the programs developed and to work on the long-term question of gender-equity in the engineering profession."[81] The Chair renewal application was successful, and Monique's term as Chair for Women in Engineering was extended until 1999, with continued industry support from Northern Telecom for a total of $900,000—between NSERC and Northern Telecom—over five years.[82]

The Second Term (1994–1997)

As her second term as Chair for Women in Engineering began, Monique started to take the Chair program in some new directions. One change was a decrease in classroom visits as the video *Engineering: Design Tomorrow's World* began to fulfill part of that role: depicting engineering as a caring profession and dismantling some of the stereotypes that surround "both the profession and the role of women in it."[83] A French language version of the video was also created, in order to reach a larger proportion of the Canadian population. In 1994 the Chair introduced Parent Sensitization Sessions, funded through the Canada/New Brunswick Cooperation Agreement on Entrepreneurship and Human Resource Development grant, to be launched in 1995. These sessions would aim to educate parents about career options for their children, and in particular, explain to parents why engineering and science are positive options for girls. Booklets as well as workshops were to be distributed and held throughout New Brunswick.[84] The Chair had also requested statistics from

[80] Monique Frize, "1993 Annual Report," pg. 2, 1993, 10–23, Box: NSERC UNB Chair 1, Folder: 1993 Annual Report, Monique Frize Fonds, University of Ottawa Archives and Special Collections, Ottawa, Ontario, Canada.

[81] Monique Frize, "Proposal to Northern Telecom and NSERC by the University of New Brunswick for support of a second term for the Chair for Women in Engineering," pg. 2, 1993,10–23, Box: NSERC UNB Chair 1, Folder: Proposal to Northern Telecom and NSERC by the University of New Brunswick for support of a second term for the Chair in Women in Engineering, Monique Frize Fonds, University of Ottawa Archives and Special Collections, Ottawa, Ontario, Canada.

[82] Monique Frize, "1993 Annual Report," pg. 6, 1993, 10–23, Box: NSERC UNB Chair 1, Folder: 1993 Annual Report, Monique Frize Fonds, University of Ottawa Archives and Special Collections, Ottawa, Ontario, Canada.

[83] Monique Frize, "1994 Annual Report," pg. 8, 1994, 10–23, Box: NSERC UNB Chair 1, Folder: 1994 Annual Report, Monique Frize Fonds, University of Ottawa Archives and Special Collections, Ottawa, Ontario, Canada.

[84] Ibid, pg. 25.

UNB concerning the enrollment and retention of engineering students by year for several years. This did not come about until 1994 as UNB was in the process of updating their computer systems to allow this type of data collection.[85]

Monique had also organized a special leave as Chairholder to take up the position as visiting professor at the University of British Columbia (UBC) from January to April of 1995, with the hopes that the exposure to Western Canada would enrich "the Chair Holder's perspectives and expose interested parties in British Columbia to the experience and skills of the Chair Holder."[86] Monique was invited to UBC as a visiting professor after meeting with Dr. Axel Meisen, then Dean of Applied Science at UBC, in November of 1994 at a conference in Vancouver.[87] She felt that a change of pace and meeting new people would help the Chair program advance, but it would also "recharge her batteries." Unfortunately, at UNB, not all professors in engineering were supportive of the Chair program, and Monique felt there was a general lack of interest within the faculty about the Chair's activities.[88] Axel, on the other hand, was very interested in the Chair program and, while there, met with Monique every two weeks for a debriefing. Monique found this to be quite refreshing and motivating.[89] While the program was supposed to cover the whole country, being located in one province limited the reach that Monique could have in Western Canada, and outside of New Brunswick in general. One of the goals of the visiting professorship at UBC was to "increase the awareness of students, faculty members and local engineers on issues and problems that the traditional culture has created for women who have chosen this field of endeavour."[90] While at UBC, Monique had the opportunity to network with students, engineers, faculty, and stakeholders and was able to deliver some of that information. Two women engineers Monique was able to meet with and discuss what the Chair program was about were Dr. Elizabeth Croft, an assistant professor at UBC, and Dr. Elizabeth Cannon, an assistant professor at the University of Calgary. Both of these women would later become regional Chairs for Women in Science and Engineering.[91]

While at UBC, Monique took on a speaking tour in Asia. Axel helped in the organization of this tour, along with the Association of Southeast Asian Nations (ASEAN) Centre in Singapore. The goal of this tour was to exchange information and ideas with "women scientists, engineers, professors and students" about how to increase the participation of women in these fields. Monique visited and spoke in three cities: Hong Kong, Kuala Lumpur, and Bangkok.[92] In her report on the Pan-

[85] Ibid, pg. 9.

[86] Ibid, pg. 28.

[87] Monique Frize, interview by Hannah Elizabeth Young, telephone interview, May 7, 2018.

[88] Ibid.

[89] Ibid.

[90] Monique Frize, "Appendix F: Chair Reports from British Columbia," pg.9, 1995, 10–23, Box: NSERC UNB Chair 1, Folder: 1995 Annual Report, Monique Frize Fonds, University of Ottawa Archives and Special Collections, Ottawa, Ontario, Canada.

[91] Ibid pg. 19, 30–3.1.

[92] Ibid pg. 2.

Asian tour, Monique wrote positively about the eagerness and hope of the women engineers she met and suggested that "perhaps we can learn from the positive aspects of each culture and that progress will only come about when the political will of those in power support it." Until then, she suggested that education would be paramount, and that women need to be empowered to use it "in the very best way they can," through building their own businesses, and establishing economic power by supporting one another and networking."[93]

At the completion of her visiting professorship at UBC, Monique wrote a report detailing her activities as Chair, as well as ten recommendations for how UBC can become more women-friendly. The recommendations included the organization of "focus groups for people to discuss and address some of their concerns, in a non-threatening environment," as well as for UBC to increase the percentage of women teachers and researchers at the university.[94] The last recommendation made, however, was the most extensive:

> It seems that the University of British Columbia and the other universities in this province would benefit greatly from having a Chair on Women in Engineering and Science, to serve all four Institutions. The emphasis should definitely be on Engineering (2/3) with some component reserved for Science, Forestry, Computer Science, as the issues are similar and similar strategies and information can be shared between all these non-traditional fields. The Chair should have a Junior Chair component to provide the complementary support needed. For example, the Junior Chair could be from a field not in science and engineering but held by a person with knowledge on women in these fields. An alternative idea is to hire a young woman engineer, on a part-time basis, as a Junior Chair, if this is appropriate for the circumstances and design of the work plan. These ideas need to be elaborated and I would be pleased to assist UBC in developing this concept further.[95]

During her time at UBC, it had become clear to Monique that there was much to be done to improve the climate for, and the representation of, women in engineering throughout the country, and it was too big a job for one National Chair. It was not just engineering either, as women in other science, technology, and mathematics fields faced some of the same issues, yet there was no Chair attempting to solve them. After her return to UNB, Monique continued to pursue the idea of multiple Chairs, although it would not be realized for another year.

Following the release of the *More Than Just Numbers* report, Monique wrote an article, titled "Reflection on the Engineering Profession: Is it becoming friendlier for women?" in the Canadian Society for Mechanical Engineering (CSME) Bulletin of 1993. If she did acknowledge that the engineering profession was becoming friendlier for women, she noted that it appeared that changes were being made to varying degrees across the country a year after the release of the report. Thus, Monique decided to organize a follow-up conference in Fredericton in the spring of

[93] Monique Frize, "Women in Engineering: A West Coast Experience," pg. 5, 1995, 10–23, Box: NSERC UNB Chair 1, Folder: Women in Engineering: A West Coast Experience—Report, Monique Frize Fonds, University of Ottawa Archives and Special Collections, Ottawa, Ontario, Canada.

[94] Ibid pg. 11.

[95] Ibid pg. 14.

1995 with the former CCWE and other key persons. The goal was to touch base with the engineering community in Canada and to see how change had been enacted, what was effective, and what was not. In spite of the Chair's efforts to recruit elementary and secondary school teachers, as well as to invite engineering deans across the country, only one teacher and approximately one third of engineering deans attended. Despite this, representatives from industry, including Gary Donahue from Nortel (Northern Telecom had changed their name to Nortel by this point), Peter Morand, president of NSERC, representatives from the Division for the Advancement of Women Engineers and Geoscientists in British Columbia (DAWEG) and Canadian Coalition for Women in Science, Engineering, Technology and the Trades (CCWESTT), as well as Ursula Franklin spoke, showing support for change within all aspects of the engineering community.[96] Additionally, the overwhelming consensus of the conference was that changes had been made to make engineering more women-friendly: "while the recommendations were not implemented in their entirety, or by every employer, changes were reported in industry that are encouraging."[97]

Throughout the conference, the attendees were given the opportunity through discussion groups to share their experiences and provide suggestions to build upon the recommendations of the 1992 report. One important recommendation from a participant was that boys should be included in target groups because they needed to see more women in engineering and in other professions. Another attendee argued that programs targeting parents helped them foster more appreciation for nontraditional career choices. A third recommendation advanced that proactively searching for women faculty and hiring them is a solution which will help to bring positive change to many of the other issues mentioned.[98] Other recommendations focused on diversity training, and increased training for women to better prepare them for male-dominated environments. The importance of valuing feminine attributes equally to those considered masculine was also discussed.[99] One attendee declared that "with the exception of isolation, women in engineering are facing the same issues as women in all professions. This highlights the need to network and find solutions with other women, for all women."[100] In hindsight, the *More Than Just Numbers* update conference brought together women engineers, male engineers, stakeholders,

[96] Frize, *A Woman in Engineering: Memoirs of a Trailblazer*, Ottawa, University of Ottawa Press, 2019, p. 127–128.

[97] Monique Frize, "1995 Annual Report," pg. 13, 1994, 10–23, Box: NSERC UNB Chair 1, Folder: 1995 Annual Report, Monique Frize Fonds, University of Ottawa Archives and Special Collections, Ottawa, Ontario, Canada.

[98] Monique Frize, "Appendix V: More Than Just Numbers 1995 Conference Discussion Group Outcomes," pg. 37, 1995, 10–23, Box: NSERC UNB Chair 1, Folder: 1995 Annual Report, Monique Frize Fonds, University of Ottawa Archives and Special Collections, Ottawa, Ontario, Canada.

[99] Ibid.

[100] Ibid.

and academia to discuss how to make the environment more women-friendly, and better for all. Once again, despite change having occurred, it was clear that the work of the Chair for Women in Engineering was still very much needed.

Five New Chairs (1996–1997)

In December 1994, NSERC began a restructuring of its granting programs. The funding agency recognized that they were not doing all they could to encourage the participation of women in their programs, and in order to solve this, they established a Task Force on Women in Science and Engineering in the early months of 1996. The Task Force was responsible for producing a report that would advise the NSERC Council on "how to create an environment that facilitates and encourages greater participation of women in science and engineering research in Canada, particularly through NSERC's core programs."[101] The 13-member Task Force was chaired by Monique and had members from several universities including the University of Calgary and the University of Toronto, as well as industry, and from various fields including Geosciences, Engineering, and Zoology. Dr. Claire Deschênes, a mechanical engineering professor from Université Laval who would later become Chair for Women in Science and Engineering for the Quebec region, was also on the Task Force.[102]

The staff at NSERC collected statistics about researchers receiving grants from NSERC to inform the Task Force, who came up with 21 recommendations in their final report, titled *Towards Building a New Scientific and Engineering Culture in Canada*.[103] The Task Force submitted its report to NSERC in June 1996, and, one month later, the Report of the Task Force on Women in Science and Engineering to the Natural Sciences and Engineering Research Council was released to the public with the support of the President of NSERC, Thomas A. Brzustowski: "In my view, Dr. Monique Frize and the Task Force are to be congratulated for producing an excellent and important report, and I commend it to your serious attention."[104] The 21 recommendations ranged from the re-instatement of a WFA (Women Faculty Awards)-type program at NSERC, which had been shut down in 1994, to the adoption of a policy statement on Women in Science and Engineering, to the

[101] Thomas A. Brzustowski, "Message from the President," pg. 2, 1996, 10–23, Box: NSERC UNB Chair 1, Folder: Report of the Task Force on Women in Science and Engineering to the Natural Sciences and Engineering Research Council, Monique Frize Fonds, University of Ottawa Archives and Special Collections, Ottawa, Ontario, Canada.

[102] Ibid.

[103] Thomas A. Brzustowski, "Message from the President," pg. 2, 1996, 10–23, Box: NSERC UNB Chair 1, Folder: Report of the Task Force on Women in Science and Engineering to the Natural Sciences and Engineering Research Council, Monique Frize Fonds, University of Ottawa Archives and Special Collections, Ottawa, Ontario, Canada.

[104] Ibid.

identification by NSERC of women faculty candidates for university positions. One recommendation, however, initially made by Monique, was the introduction of multiple regional Chairs for Women in Science and Engineering.[105]

When the Task Force introduced the report to the NSERC Council in June 1996, the Council "endorsed it enthusiastically, and asked that it be distributed to all NSERC committees, university presidents and senior administrators, and that it also be made available on NSERC's World Wide Web site to ensure widespread dissemination."[106] The Council also adopted this policy statement:

> NSERC believes that women should have every opportunity to participate fully in science and engineering and strongly encourages women to consider entering careers in these fields. The Council's programs and peer review system are designed and operated to ensure fair treatment of all applicants regardless of gender. NSERC recognizes that barriers still exist to the full participation of women in science and engineering careers. It is committed to identifying and working to eliminate any barriers that may exist within its own programs and processes and encourages others to do the same. In addition, NSERC endorses the use of special initiatives when appropriate to encourage and assist women in reaching full participation in all disciplines.[107]

The Council also worked quickly to implement the majority of the 21 recommendations, including the recommendation for multiple regional Chairs. They agreed to create five Chairs in addition to the one in existence, and each of the Chairs would be Chairs for Women in Science and Engineering, responsible for a particular geographical region in Canada. Like the National Chair, each would be a joint NSERC-industry sponsored position for a five-year term with the option for a one-time renewal. NSERC began planning immediately for a competition to begin in the fall of 1996.[108] The five regions would be West Coast and Yukon, Prairies, Ontario, Quebec, and Atlantic.[109] When asked about the decision to recommend a regional

---

[105] NSERC Task Force for Women in Science and Engineering, "Report on Action Taken to Date on Recommendations of the Task Force on Women in Science and Engineering," pg. iv, 1996, 10–23, Box: NSERC UNB Chair 1, Folder: Report of the Task Force on Women in Science and Engineering to the Natural Sciences and Engineering Research Council, Monique Frize Fonds, University of Ottawa Archives and Special Collections, Ottawa, Ontario, Canada.

[106] Thomas A. Brzustowski, "Message from the President," pg. 2, 1996, 10–23, Box: NSERC UNB Chair 1, Folder: Report of the Task Force on Women in Science and Engineering to the Natural Sciences and Engineering Research Council, Monique Frize Fonds, University of Ottawa Archives and Special Collections, Ottawa, Ontario, Canada.

[107] Ibid.

[108] NSERC Task Force for Women in Science and Engineering, "Report on Action Taken to Date on Recommendations of the Task Force on Women in Science and Engineering," pg. iv, 1996, 10–23, Box: NSERC UNB Chair 1, Folder: Report of the Task Force on Women in Science and Engineering to the Natural Sciences and Engineering Research Council, Monique Frize Fonds, University of Ottawa Archives and Special Collections, Ottawa, Ontario, Canada.

[109] Five New Chairs Press Release, pg. 2, 1996, 10–23, Box: NSERC UNB Chair 1, Folder: 1996 Annual Report, Monique Frize Fonds, University of Ottawa Archives and Special Collections, Ottawa, Ontario, Canada.

Chair system, Monique responded that she simply needed help.[110] The Task Force report created lasting change within NSERC, and according to Monique, "increased the awareness of the officers at NSERC on the need to be proactive, to encourage more women."[111] The creation of the five regional Chairs for Women in Science and Engineering program was the first of its kind in the world and a model that would be used in following years by the United Nations Educational, Scientific and Cultural Organization (UNESCO), as they created Chairs in different countries throughout the globe.[112]

During her time as a visiting professor at UBC and NSERC's launch for the five new Chairs, Monique applied for a sabbatical from UNB. Due to issues with members of the Faculty of Engineering at UNB, Monique felt that that environment was no longer conducive to the Chair program.[113] During her sabbatical, which took place over 1996 and the beginning of 1997, Monique returned to Ottawa, where she had grown up and attended university, and split her time between Carleton University and the University of Ottawa.[114] When the competition for the five regional Chairs launched on October 5, 1996, the two universities applied for the Ontario Chair, with Monique as incumbent, and with continued industry support from Nortel; in March 1997, Monique was selected for the position.[115]

***

When looking back at her seven years as NSERC-Nortel Chair for Women in Engineering in her final report, Monique said "while I was the only Chair of this kind in Canada, the effort required was enormous (and I must admit, exhausting), if the work was to be done properly and in a timely manner."[116] There were many reasons to be optimistic, however, and Monique looked forward to working in partnership with four other Chairs: "the sharing of the responsibilities with four other women will allow us to build many new aspects on what I believe is already a strong foundation left after seven years of efforts."[117] When Monique began as National

---

[110] Monique Frize, interview by Hannah Elizabeth Young, telephone interview, May 7, 2018.

[111] Ibid.

[112] Ibid.

[113] Ibid.

[114] Monique Frize, "1997 Final Report," pg. 4, 1997, 10–23, Box: NSERC UNB Chair 1, Folder: 1997 Final Report, Monique Frize Fonds, University of Ottawa Archives and Special Collections, Ottawa, Ontario, Canada.

[115] 22 Five New Chairs:

[116] Monique Frize, "1997 Final Report," pg. 10, 1997, 10–23, Box: NSERC UNB Chair 1, Folder: 1997 Final Report, Monique Frize Fonds, University of Ottawa Archives and Special Collections, Ottawa, Ontario, Canada.

Five New Chairs Press Release, pg. 1, 1996, 10–23, Box: NSERC UNB Chair 1, Folder: 1996 Annual Report, Monique Frize Fonds, University of Ottawa Archives and Special Collections, Ottawa, Ontario, Canada.

[117] Monique Frize, "1997 Final Report," pg. 10, 1997, 10–23, Box: NSERC UNB Chair 1, Folder: 1997 Final Report, Monique Frize Fonds, University of Ottawa Archives and Special Collections, Ottawa, Ontario, Canada.

Chair and was travelling around the country giving speeches—which she continued to do throughout her time as regional Chair—she would tell the women she spoke to that "they didn't have to be like the guys…they can be women; they can be themselves."[118] This was a stark contrast to her own experience as an undergraduate engineering student in the 1960s, when she had to do her utmost to blend in. Having a female role model, even if it is just one, is significant, and can make any person feel less alone when they walk an unconventional path, and that is what Monique succeeded in doing for many women throughout her seven years as NSERC-Northern Telecom/Nortel Chair for Women in Engineering.

[118] Monique Frize, interview by Hannah Elizabeth Young, telephone interview, May 7, 2018.

Chapter 2
The First Regional Chairs for Women in Science and Engineering (1997–2003)

The submission of the Task Force's final report in June 1996 launched a new phase for the National Chair program. The final report argued that the extent of the task was just too considerable to be accomplished by a single person for the entire nation. The report not only suggested to add positions of this type in several geographical areas of the country, but it also proposed to enlarge the mandate from engineering only to include science and other related fields. Recognizing that supporting such a program would come with a sizable cost, it recommended to work in partnership with private or public sector sponsors. Since the budget for Industrial Research Chairs had predominantly served men in the past, it was argued that Natural Sciences and Engineering Research Council of Canada (NSERC) could use a portion of that budget to expand the program to include science and other related fields and provide additional role models for young women considering these disciplines.

Following the release of the Task Force report, and NSERC's enthusiastic endorsement, the council agreed to fund, through its Industrial Chair budget, five Chairs for Women in Science and Engineering (WISE) responsible for a particular geographical region in Canada (West Coast and Yukon, Prairies, Ontario, Quebec, and the Atlantic). Like the National Chair, each would be a joint NSERC-industry-university sponsored position for a five-year term with the option for a one-time renewal. Each of these regional Chairs would be assessed prior to the termination of the first mandate, and renewal was possible for a second three-year term if the evaluation was satisfactory. NSERC began planning immediately for a hiring competition to begin in the fall of 1996.[1]

---

[1] NSERC Task Force for Women in Science and Engineering, "Report on Action Taken to Date on Recommendations of the Task Force on Women in Science and Engineering," p. iv, 1996, 10–23, Box: NSERC UNB Chair 1, Folder: Report of the Task Force on Women in Science and Engineering to the Natural Sciences and Engineering Research Council, Monique Frize Fonds, University of Ottawa Archives and Special Collections. See also Monique Frize, *Towards Building a New*

C. D'Amours et al., *The Chairs for Women in Science and Engineering Program*, Women in Engineering and Science, https://doi.org/10.1007/978-3-031-54085-1_2

On April 18, 1997, the Government of Canada announced the details of the regional Chairs for Women in Science and Engineering (CWSE) program, a "$2.6 million initiative to promote women in science and engineering," to be launched in July 1997. The press release recalled the importance of science and engineering educated workers to Canadian prosperity in a knowledge-based economy and the need to showcase more role models of women working in these fields to attract more young women in this career path: "science and engineering are key to job creation and economic growth in the new economy. These Chair holders are highly regarded researchers who will provide outstanding role models for young women."[2] The document defined the role of the Chairholders, which remained much the same as the role of the National Chair: they would spend half of their time as a role model through fulfilling academic and scholarly functions and spent the other half on activities that would lead to increased participation of women in their professions. Except there was one important change, the Chair program expanded now to include women in sciences, reflecting NSERC's mandate and scope.[3]

The press release also presented the five successful applicants to the competition: as Atlantic Chair, Dr. Mary Williams, who was sponsored by Petro-Canada and would join the Faculty of Engineering and Applied Science, as well as the Faculty of Science at Memorial University; as Quebec Chair, Dr. Claire Deschênes, who would be sponsored by Alcan Aluminium Limitée and would continue working in the Department of Mechanical Engineering at Université Laval; as Ontario Chair, Dr. Monique Frize, who would continue to be sponsored by Nortel, and would join the Faculty of Engineering at both the University of Ottawa and Carleton University; as Prairie Chair, Dr. Elizabeth Cannon, who would also be sponsored by Petro-Canada and would remain in the Department of Geomatic Engineering at the University of Calgary; and lastly, as the Western Canada Chair, Dr. Maria Klawe, sponsored by IBM Pacific Development Team, who would continue at the University of British Columbia in the Department of Computer Science.[4]

Scientific and Engineering Culture in Canada. Report of the Task Force on Women in Science and Engineering, Ottawa, NSERC, 1996, p. 21–22, 10–23, Box: NSERC UNB Chair 1, Folder: Report of the Task Force on Women in Science and Engineering to the Natural Sciences and Engineering Research Council, Monique Frize Fonds, University of Ottawa Archives and Special Collections. See also Monique Frize, Claire Deschênes, Elizabeth Cannon, Mary Williams, and Maria Klawe, "A Unique National Project to Increase the Participation of Women in Science and Engineering from Canada", *Nature*, 23 September 1999, accessed November 28, 2018: https://www.nature.com/nature/debates/women/women_10.html

[2] Five New Chairs Press Release, p. 1, 1997, 10–123, Box: NSERC Chairs, Folder: News Release Federal, Monique Frize Fonds, University of Ottawa Archives and Special Collections.

[3] Five New Chairs Press Release, p. 1, 1997, 10–123, Box: NSERC Chairs, Folder: News Release Federal, Monique Frize Fonds, University of Ottawa Archives and Special Collections and Anne Millar, "A National Mandate to Solve the 'Problem' of Women in Science and Engineering: The Federal Government and the Creation of Chairs for Women in Science and Engineering", paper presented to the *Canadian Historical Association – 91st Annual Meeting*, Waterloo, May 28, 2012.

[4] Five New Chairs Press Release, p. 1–2, 1997, 10–123, Box: NSERC Chairs, Folder: News Release Federal, Monique Frize Fonds, University of Ottawa Archives and Special Collections.

A Solid National Initiative

In August 1997, Monique Frize invited the four other Chairholders to meet in Ottawa to ensure a smooth transition between one National Chair to five regional Chairs and to secure effective cooperation between all of them. This first meeting proved to be instrumental to the formation of a strong national network to support women in science and engineering. Although the five Chairs all had different backgrounds in terms of culture, education, or leadership experience, there was a strong camaraderie between all of them from the start. Certainly, this personal and professional solidarity eased the transition. This helped them develop and implement a communication and networking plan from the start, to ensure regional and national impacts on opportunities for women in science and engineering[5] (Fig. 2.1).

To reinforce the bond, the Chairs chose to meet at least once per year, the meeting being held in a different location every time, in order to visit each region at least once during the five-year term. The Chairs also decided to use these meetings as a means to interact with the larger community, inside and outside the academic environment. Usually, the Chairs would meet with women faculty members in science and engineering, researchers working around gender and science, and also with the local Chair's advisory council, when such a body existed. Doing so supported the effective communication of a national vision for the five regional Chairs while reinforcing to the local community the work being done by their regional Chair.[6] In total, eight meetings were held during their first term. Between these meetings, several teleconferences were held, and, in some cases, a subset of the Chairs held informal meetings when they had the opportunity to be together.[7]

The five regional Chairs were not only anxious to build meaningful relationships between themselves, they were also eager to project a strong national message. In order to do that, it was decided that each Chairholder would act as "coordinator of the Chairs" for a one-year term to (1) coordinate and host a meeting; (2) arrange

[5] Monique Frize, Interview with Hannah Elizabeth Young, telephone interview, May 8, 2018; Maria Klawe, Interview with Hannah Elizabeth Young, telephone interview, September 4, 2018; and Barbara Muir, NSERC, "The NSERC/Industry Chairs for Women in Science and Engineering: A National Program in Canadian Universities", paper presented to 12th International Conference of Women Engineers and Scientists, Ottawa, July 2002, accessed November 28, 2018, https://www.mun.ca/cwse/icwes_cwse5.pdf and Anne Millar, "A National Mandate to Solve the 'Problem' of Women in Science and Engineering: The Federal Government and the Creation of Chairs for Women in Science and Engineering", paper presented to the *Canadian Historical Association – 91st Annual Meeting*, Waterloo, May 28, 2012.

[6] Barbara Muir, NSERC, "The NSERC/Industry Chairs for Women in Science and Engineering: A National Program in Canadian Universities", paper presented to 12th International Conference of Women Engineers and Scientists, Ottawa, July 2002, accessed November 28, 2018, https://www.mun.ca/cwse/icwes_cwse5.pdf

[7] F.M. Williams et C.J. Emerson, *48 month (sic) progress report submitted to Natural Sciences and Engineering Research Council of Canada*, mars 2001, p. 15; Elizabeth Cannon, *Progress Report #2: NSERC/Petro-Canada Chair for Women in Science and Engineering (Prairies Region)*, Calgary, University of Calgary, août 2001, p. 14.

Fig. 2.1 The five first-generation Chairs for Women in Science and Engineering meeting in Ottawa. From left to right, F. Mary Williams (Atlantic Chair), Elizabeth Cannon (Prairie Chair), Claire Deschênes (Quebec Chair), Maria Klawe (British Columbia & Yukon Chair), Monique Frize (Ontario Chair)

teleconferences and broadcast email messages; (3) coordinate a national newsletter; and (4) serve as a member of the board of the Canadian Coalition of Women in Engineering, Science, and Technology (CCWEST). In addition, they made sure to meet with leaders in the science and engineering community, including the National Committee of Deans of Science in May 1999, the National Committee of Deans of Engineering and Applied Science in May 1999 and March 2001, the Secretary of State for Science and Technology, and the President of NSERC. The Chairs also responded collectively to a number of media reports on women in science and engineering issues so that factual information was communicated to stakeholders and the public at large. For example, the five NSERC Chairs reacted to the negative letter published by Doreen Kimura, the president of the Society for Academic Freedom and Scholarship, in the *National Post* in January 2000 on the NSERC's University Faculty Awards, a program which selectively supported new women faculty. Furthermore, they took active participation in a number of national and international organizations such as INWES, Asia-Pacific Economic Cooperation (APEC), CCPE, and the United Nations Educational, Scientific and Cultural Organization (UNESCO).[8]

---

[8] Barbara Muir, NSERC, "The NSERC/Industry Chairs for Women in Science and Engineering: A National Program in Canadian Universities", paper presented to 12th International Conference of Women Engineers and Scientists, Ottawa, July 2002, accessed November 28, 2018, https://www.mun.ca/cwse/icwes_cwse5.pdf; Archives Monique Frize, uOttawa, NSERC/Nortel 48 Month Report – September 2001 and Atlantic Chair on Women in Science and Engineering, "NSERC

To strengthen their action throughout Canada, the Chairs were also involved in the development of several national initiatives, including a national newsletter which was published yearly and was disseminated through the Chairs' websites, and the creation of a bibliographic database containing reference material on gender and science and engineering, accessible through some of the Chairs' websites. Along with two national conferences on women in science and engineering, the Chairs also organized international events during their five-year term (in Vancouver in 1998, St. John's in 2000, and Ottawa in 2002). Although the conferences were hosted by individual Chairs, they truly became national events due in large part to the efforts of the five Chairs in identifying contributions and encouraging participation from their respective regions. Through these initiatives, the five Chairs provided a strong voice on women in science and engineering issues from coast to coast.[9]

Besides forming a solid network and projecting a national message, the five regional Chairs also worked closely together to ensure an effective sharing of strategies, experiences, and information. They met in Calgary in December 1997 to elaborate a strategy at both the national and regional levels, the results of which were presented in a joint paper at the Engineering Foundation Conference on Women in Engineering in July 1998. The development of a *National Action Plan* was thus crucial to effectively promoting a common vision and course of action given the broad and ambitious objectives of the program, and considerable previous activity by other agencies.[10] With this strategy, the five regional Chairs effectively subscribed to the paradigm shift characteristic of the first Chair, the CCWE, and the *More than Just Numbers* report: they recognized that the number of women entering engineering and science could only grow steadily if some fundamental changes were made to the culture and the education process. Success was more than just attracting greater numbers of women into these professions. It was also about ensur-

---

University Faculty Awards", 2000, accessed January 20, 2019, http://www.mun.ca/cwse/events_news.html

[9] F.M. Williams et C.J. Emerson, *48 month (sic) progress report submitted to Natural Sciences and Engineering Research Council of Canada*, mars 2001, p. 15; Elizabeth Cannon, *Progress Report #2: NSERC/Petro-Canada Chair for Women in Science and Engineering (Prairies Region)*, Calgary, University of Calgary, août 2001, p. 16–17 and Barbara Muir, NSERC, "The NSERC/Industry Chairs for Women in Science and Engineering: A National Program in Canadian Universities", paper presented to 12th International Conference of Women Engineers and Scientists, Ottawa, July 2002, accessed November 28, 2018, https://www.mun.ca/cwse/icwes_cwse5.pdf

[10] Elizabeth Cannon, *Progress Report #2: NSERC/Petro-Canada Chair for Women in Science and Engineering (Prairies Region)*, Calgary, University of Calgary, août 2001, Appendix VI: National Action Plan, p. 66; Monique Frize, Claire Deschênes, Elizabeth Cannon, Mary Williams, and Marie Klawe, "A Unique National Project to Increase the Participation of Women in Science and Engineering" in Barbara Bogue, Priscilla Guthrie, Barbara Lazarus, and Steve Hadden, *Tackling the Engineering Resources Shortage: Creating New Paradigms for Developing and Retaining Women Engineers*, Bellingham, International Society for Optical Engineering, 1999, p. 83 and NSERC Regional Chairs for Women in Science and Engineering, *Newsletter*, vol. 2, no 1 (mars 1999), p. 1, Monique Frize Fonds, University of Ottawa Archives and Special Collections.

ing that women have positive experiences in these learning and working environments. Thus, in addition to tending to general issues of attraction and retention, the five regional Chairs were also engaged in a closer analysis of the curriculum, structures, and conventions that were mitigating the full participation of women in engineering and science. This orientation explains their focus on institutional transformation and their demands for government-backed programs at the national level. This emphasis on cultural influences and stereotypes also reveals how much they valued feminist scholarship and were thus dependent on interdisciplinary collaborations to find solutions to gender issues in engineering and science disciplines.[11]

Yet, if the regional Chairs' programs had the same ultimate objectives and were following the same strategic plan, all five were very much different in content, audience, and scope as we shall see in the next sections. Indeed, the objectives were so extensive that it was impossible for one Chairholder to do everything all at once in her own region. Each Chair had to analyze for herself where she would focus her effort according to her identity, her own strengths, what had already been done inside or outside her jurisdiction, and where she could add something or assist with other organizations which were already working in this space. Thus, factors like personality, geography, politics, and culture all influenced the way each Chairholder envisioned her own program, ultimately generating five very different agendas oriented to one single goal: to make science and engineering much more attractive to girls and women.[12]

---

[11] Monique Frize, Claire Deschênes, Elizabeth Cannon, Mary Williams and Maria Klawe, "A Unique National Project to Increase the Participation of Women in Science and Engineering from Canada", *Nature Debates*, 23 September 1999, https://www.nature.com/nature/debates/women/women_10.html and Anne Millar, "A National Mandate to Solve the 'Problem' of Women in Science and Engineering: The Federal Government and the Creation of Chairs for Women in Science and Engineering", paper presented to the *Canadian Historical Association – 91st Annual Meeting*, Waterloo, May 28, 2012; Monique Frize and Jeanne Inch, "Women in Engineering: More than Just Numbers" in Cathy M. Deno (ed.), *Proceedings of the 1992 Women in Engineering Conference: Increasing Enrollment and Retention*, West Lafayette, WEPAN, 1992, p. 5, accessed November 20, 2018, https://journals.psu.edu/wepan/article/view/57600/57288 and Monique Frize, *The Bold and the Brave: A History of Women in Science and Engineering*, Ottawa, University of Ottawa Press, 2009, p. 161.

[12] Barbara Muir, NSERC, "The NSERC/Industry Chairs for Women in Science and Engineering: A National Program in Canadian Universities", paper presented to 12th International Conference of Women Engineers and Scientists, Ottawa, July 2002, accessed November 28, 2018, https://www.mun.ca/cwse/icwes_cwse5.pdf; Anne Millar, "A National Mandate to Solve the 'Problem' of Women in Science and Engineering: The Federal Government and the Creation of Chairs for Women in Science and Engineering", paper presented to the *Canadian Historical Association – 91st Annual Meeting*, Waterloo, May 28, 2012; Carolyn Emerson, Interview with Hannah Elizabeth Young, telephone interview, January 3, 2018; Elizabeth Cannon, Interview with Hannah Elizabeth Young, telephone interview, May 3, 2018; Maria Klawe, Interview with Hannah Elizabeth Young, telephone interview, September 4, 2018; Claire Deschênes, Interview with Hannah Elizabeth Young, telephone interview, March 13, 2018; and Elizabeth Cannon, *Progress Report #2: NSERC/Petro-Canada Chair for Women in Science and Engineering (Prairies Region)*, Calgary, University of Calgary, août 2001, Appendix VI: National Action Plan, p. 66.

The Ontario Chair: Monique Frize (1997–2004)

Being appointed as the new Chair for Women in Science and Engineering for the Ontario region allowed Monique Frize to finally come back permanently to Ottawa. After her hardship at the University of New Brunswick (UNB), it must have felt like a relief. On the professional side, she was coming back to her *alma mater*, the University of Ottawa, which jointly hosted the Chair with Carleton University. As the University of Ottawa's first female engineering graduate, the wheel had come full circle. The presence of two major health institutes, the University of Ottawa Heart Institute and the University of Ottawa Eye Institute of the Ottawa Hospital, also allowed her to get closer to her own interests in the biomedical community.[13] Her long-term goal was still to increase the participation of women in all aspects of engineering—and now science—careers. But her responsibilities now covered only a province instead of a whole country. This was a nice change of pace for her: "Of course Ontario had more universities than anywhere else, but for me it was so much less travel than before."[14]

In the national capital, Monique initially focused on expanding the outreach and retention activities in the Ottawa/Carleton area by strengthening the existing groups of volunteers dedicated to these tasks. The first project she set her eyes on was the *Pathmakers Program*. This partnership between the University of Ottawa, Carleton

[13] Carleton University and the University of Ottawa Press release, "New Joint Chair for Women in Engineering and Science will 'bring engineering and science closer to people'", p. 1, 1997, 10–123, Box: NSERC Chairs, Folder: NSERC Chairs - News release Carleton UO, Monique Frize Fonds, University of Ottawa Archives and Special Collections and Monique Frize, Interview with Hannah Elizabeth Young, telephone interview, May 7, 2018.

[14] Monique Frize, Interview with Hannah Elizabeth Young, telephone interview, May 7, 2018.

University, Algonquin College, and the four school boards in the Ottawa-Carleton district was originally initiated by the Ontario Women's Directorate in 1986. The first goal of the program was to encourage young people, especially girls and young women, to be informed about engineering, science, or technology opportunities and pursue a career in these fields. A second objective was to develop communication between role models and prospective students in both elementary and secondary schools. Through this program, volunteer female students enrolled in science and engineering undergraduate and graduate degree programs were invited by schools to make presentations and conduct hands-on activities with boys and girls. This could take the form of experiments, demonstrations, and career information geared to the age group concerned.[15]

After Monique's return to Ottawa in 1997, she began not only coordinating the activities of the program but also its fundraising and promotion. In 1997–1998, the program attracted more than 250 volunteers, a fourfold increase over the previous years. In 2000, 150 volunteers gave more than 80 school presentations. The direct results were impressive. Between 1996 and 2000, enrollment of women in undergraduate engineering programs at Carleton University and the University of Ottawa went from 12% to around 24%, this increase happening even in traditionally male fields like electrical and mechanical engineering.[16] The *Pathmakers Program* was so successful that it won the 1999 Outreach Award from the 1989 Canadian Memorial Foundation.[17]

Monique also took active participation in the Ottawa-Carleton WISE Chapter as the new president of this section. In the late 1990s, the local organization was in a dormant state, not having held any meetings in the past few years. Following a meeting with Monique and three other members of the local organization, the Ottawa-Carleton WISE Chapter was revived in November 1997. This provided new impetus for activities and programs in the Ottawa-Carleton/Hull area as these

---

[15] Monique Frize, *The Bold and the Brave: A History of Women in Science and Engineering*, Ottawa, University of Ottawa Press, 2009, p. 170; Monique Frize, Claire Deschênes, Elizabeth Cannon, Mary Williams, and Marie Klawe, "A Unique National Project to Increase the Participation of Women in Science and Engineering" in Barbara Bogue, Priscilla Guthrie, Barbara Lazarus, and Steve Hadden, *Tackling the Engineering Resources Shortage: Creating New Paradigms for Developing and Retaining Women Engineers*, Bellingham, International Society for Optical Engineering, 1999, p. 87–88 and NSERC Regional Chairs for Women in Science and Engineering, *Newsletter*, vol. 2, no 1 (mars 1999), p. 3, Monique Frize Fonds, University of Ottawa Archives and Special Collections.

[16] Monique Frize, *The Bold and the Brave: A History of Women in Science and Engineering*, Ottawa, University of Ottawa Press, 2009, p. 170 and Monique Frize, Claire Deschênes, Elizabeth Cannon, Mary Williams, and Marie Klawe, "A Unique National Project to Increase the Participation of Women in Science and Engineering" in Barbara Bogue, Priscilla Guthrie, Barbara Lazarus, and Steve Hadden, *Tackling the Engineering Resources Shortage: Creating New Paradigms for Developing and Retaining Women Engineers*, Bellingham, International Society for Optical Engineering, 1999, p. 87–88.

[17] Monique Frize, *NSERC/Nortel Women in Engineering Chair – 48 Month Report*, September 2001, p. 2, 10–123, Box: NSERC Chairs, Folder: *NSERC/Nortel Women in Engineering Chair – 48 Month Report*, Monique Frize Fonds, University of Ottawa Archives and Special Collections.

women could become mentors for the *Pathmakers Program*. Student branches were also created at the University of Ottawa and Carleton University, a valuable addition for future recruitment and retention activities of the Chair and the WISE Chapter in the capital area. In just four years, the local organization grew from a handful of participants to approximately 150 members. The revitalization of the Ottawa-Carleton WISE Chapter was also crucial for the creation of a new section in Sudbury in December 1998. Monique additionally developed the Women in Science and Engineering Network (WISEN-Ontario) with all universities in the province and other organizations, such as student leader associations and Professional Engineers Ontario. Through this network, the Ontario Chair gathered a wealth of information on the status of women in science and engineering undergraduate and graduate programs in her province. In each university, a representative was responsible for providing statistics on enrollment, notifying the network of successful strategies, and creating a node in their region if possible.[18]

Besides her involvement in these outreach and retention programs, Monique was involved in multiple research projects. Her first project focused on assessing the gender impact of science and engineering outreach activities in New Brunswick, such as the summer program *Worlds UNBound* at UNB and the Sir Robert Borden School event at Nortel. In collaboration with Ruby Heap, a historian at the University of Ottawa, she also analyzed the educational experiences of women in engineering study programs from the First World War to the end of the 20th century. Supported by a Social Sciences and Humanities Research Council grant, this project assessed the culture of engineering to identify aspects which were negative for women. Using quantitative data of women in engineering educational programs, through a questionnaire provided to a number of women engineers as well as interviews with women faculty members at the University of Ottawa, these two scholars examined the enrollment patterns in Quebec and Ontario between 1920 and 1999. Finally, Monique helped develop a new course based on their findings in collaboration with professors at the Institute of Women Studies that was first offered in the winter of 2001.[19]

[18] Monique Frize, *NSERC/Nortel Women in Engineering Chair – 1998 Annual Report*, September 1998, p. 4, 10–123, Box: NSERC Chairs, Folder: *NSERC/Nortel Women in Engineering Chair – 1998 Annual Report*, Monique Frize Fonds, University of Ottawa Archives and Special Collections; Monique Frize, Claire Deschênes, Elizabeth Cannon, Mary Williams, and Marie Klawe, "A Unique National Project to Increase the Participation of Women in Science and Engineering" in Barbara Bogue, Priscilla Guthrie, Barbara Lazarus, and Steve Hadden, *Tackling the Engineering Resources Shortage: Creating New Paradigms for Developing and Retaining Women Engineers*, Bellingham, International Society for Optical Engineering, 1999, p. 88 and Colleen M. Ennett, "Reengineering the Ottawa-Carleton WISE Chapter", *Proceedings of the 8th Annual CCWEST Conference*, St. John's, Newfoundland, 2000. Available at: http://www.mun.ca/cwse/Ennett,Colleen.pdf

[19] Monique Frize, *NSERC/Nortel Women in Engineering Chair – 1998 Annual Report*, September 1998, p. 4 and 13, 10–123, Box: NSERC Chairs, Folder: *NSERC/Nortel Women in Engineering Chair – 1998 Annual Report*, Monique Frize Fonds, University of Ottawa Archives and Special Collections; Monique Frize, *NSERC/Nortel Women in Engineering Chair – 48 Month Report*, September 2001, p. 8–9, 10–123, Box: NSERC Chairs, Folder: *NSERC/Nortel Women in*

In addition to her research activities, Monique worked closely with a number of national and international organizations. In 1996, she was asked by Industry Canada to be Chair of the Asia-Pacific Economic Cooperation (APEC) Experts Meetings on Gender, Science and Technology. In the early 1990s, the association had taken steps to positively influence women's participation in the economies of member countries. Falling under the auspices of the Industrial Science and Technology Working Group of APEC, these meetings examined gender issues in science and technology policy in knowledge-based economies. From 1996 to 2000, Monique flew twice a year to Asia to present a list of best practices for the APEC's ministers on how to increase the participation of women in science and technology. The goal was to adapt what she had found in Canada to the needs and cultures of Asian countries like China and Japan.[20]

Monique took an active role in the creation of the International Network of Women Engineers and Scientists (INWES). In July 1999, the World Conference on Science for the Twenty-First Century, organized by the United Nations Educational, Scientific and Cultural Organization (UNESCO) and the International Council for Science, adopted the Science-Agenda Framework for Action in which it was recommended to create an international network of women in Science, Technology, Engineering and Mathematics (STEM) to face the new challenges raised by increased globalization. In response to this recommendation, Monique, Claire Deschênes (a professor of mechanical engineering at Université Laval and current Quebec Chair), and Moyra McDill (a professor of mechanical and aerospace engineering at Carleton University) organized a pre-International Conference of Women Engineers and Scientists 12 (ICWES12) meeting in Ottawa in May 2001. Through UNESCO funding and support from the Canadian Commission for UNESCO, 22 ICWES representatives from ten countries and eight associations met to develop the roots of an international organization to link associations of women in science and engineering from all around the world. A year later, 551 delegates from 30 countries converged again in Ottawa for the ICWES12 conference where they unanimously voted to establish INWES. Monique and Claire were members of the interim board as president and executive secretary respectively until the first board of directors was elected in 2005, when Monique was voted in as the first official INWES president. Under her guidance, the organization grew drastically. In April 2003, INWES was incorporated in Canada as a non-profit organization and, five years later, became an official non-governmental organization partner of the operational type with

Engineering Chair – 48 Month Report, Monique Frize Fonds, University of Ottawa Archives and Special Collections. See also Monique Frize and Ruby Heap, "The Professional Education of Women Engineers in Ontario and Quebec (1920–1999): Enrolment Patterns", *Journal of the Japan Society of Mechanical Engineers*, 2001, accessed at January 26, 2019, http://www.mun.ca/cwse/Frize,Heap.pdf

[20] Monique Frize, Interview with Hannah Elizabeth Young, telephone interview, May 8, 2018, and Monique Frize, *NSERC/Nortel Women in Engineering Chair – 48 Month Report*, September 2001, p. 11, 10–123, Box: NSERC Chairs, Folder: *NSERC/Nortel Women in Engineering Chair – 48 Month Report*, Monique Frize Fonds, University of Ottawa Archives and Special Collections.

UNESCO. As of September 2018, INWES included 47 organizational members, six university/institute members, 11 honorary members, and 199 individual members, representing over 200,000 women from 45 countries around the globe.[21]

INWES
International
Network of
Women Engineers
and Scientists

The Quebec Chair: Claire Deschênes (1997–2006)

When Claire Deschênes was appointed as the NSERC/Alcan Aluminum Chair for Women in Science and Engineering in 1997, she was an associate professor in the Department of Mechanical Engineering at Université Laval. She had been hired as the first female professor of engineering at the Quebec City institution eight years earlier. Her research focused on fluid mechanics applied to hydraulic turbines. Claire founded the Laboratoire de Machines Hydrauliques (LAMH) in 1989, the only independent test laboratory in Canada specializing in research and development on hydraulic turbines using small-scale models, which would eventually become an internationally recognized center. Although hydraulic machines had always been her main interest, women's issues were never far from her mind due to her own experience as an engineering student and professor. She remembers vividly her first day at university when she found out that she was the only woman in the classroom. This came as a shock for her, being raised in a family where women were told that they could do anything they set their minds to. She was not aware that there could be any bias against women. As a professor, Claire suffered as well from the lack of credibility attributed to her gender in a male-dominate department. The Polytechnique massacre also made a profound impression on her, feeling the attack in a deeply personal way. She instantly recognized the charge against women in

[21] Monique Frize, *NSERC/Nortel Women in Engineering Chair – 48 Month Report*, September 2001, p. 12–17, 10–123, Box: NSERC Chairs, Folder: *NSERC/Nortel Women in Engineering Chair – 48 Month Report*, Monique Frize Fonds, University of Ottawa Archives and Special Collections.; Monique Frize, Interview with Hannah Elizabeth Young, telephone interview, May 8, 2018; INWES, "About", accessed January 26, 2019, http://www.inwes.org/about/ and Caroline Thoruwa, "INWES: Her story, Achievements, Challenges and Future To Build a better future worldwide", paper presented to *an Women Engineers Forum*, Mombasa, Kenya, September, 17, 2018, accessed January 19, 2019, https://www.iekenya.org/africa-engineering-conference/fawe-summit.html?download=18:caroline-thoruwa-inwes-history

STEM. These ordeals fueled her motivation to work on promoting women in these fields and her determination to become the first Quebec's Chair.[22]

The province of Quebec's distinct setting strongly influenced Claire's efforts. Even though her broad objective "to promote the fields in science and engineering to young women" and to improve employers' awareness of "both women's needs and the benefits of hiring them" was similar to her fellow regional Chairs' objectives, her program strategies needed to be adapted to Quebec's different language and culture, education system, as well as the favourable attitude of its political leader toward governmental intervention.[23] Indeed, the creation of the NSERC/Alcan Aluminum Chair happened at a pivotal moment in Quebec's history. Since the late 1970s, there has been an impulse to promote science and technology in the province. Driven by nationalist policies, the feminist movement, and the discourse surrounding the advent of the "knowledge economy," the Quebec government and several other organizations (funding agencies, regional organizations, and schools) had embarked on a vast operation to ensure the existence of an abundant high-quality workforce in science and technology. It was generally recognized in the 1990s that the creation of such a pool could not be achieved without an increased presence of women in these sectors. As the Quebec Chair for eight years, Claire served as a true catalyst for the provincial push to attract women in science and

[22] Claire Deschênes, "Honorary Doctorate Acceptance Speech", presented at the University of Ottawa president installation and fall convocation, October 23, 2016, available at: https://www.uottawa.ca/president/bio/deschenes-claire and Anne Millar and Mary Wells, "Claire Deschênes Thrives in Uncharted Waters", *CIM Magazine*, November 21, 2017, accessed January 12, 2019, http://magazine.cim.org/en/voices/claire-deschenes/

[23] Monique Frize, Claire Deschênes, Elizabeth Cannon, Mary Williams, and Marie Klawe, "A Unique National Project to Increase the Participation of Women in Science and Engineering" in Barbara Bogue, Priscilla Guthrie, Barbara Lazarus, and Steve Hadden, *Tackling the Engineering Resources Shortage: Creating New Paradigms for Developing and Retaining Women Engineers*, Bellingham, International Society for Optical Engineering, 1999, p. 88 and Paul-André Linteau, René Durocher, Jean-Claude Robert, and François Ricard, *Quebec Since 1930*, Toronto, J. Lorimier, 1989, p. 337.

engineering.[24] Two vital types of activities were undertaken to achieve the program's target: data acquisition and research as well as program development. Although not initially included in her mandate, Claire also spent a considerable amount of time working with international organizations around the world.

As part of her research, Claire's team compiled statistics to describe the evolution of the status of women in science and technology. The goal was to understand and document Quebec's status of women in these fields both at school and in the workplace, support the Chair's program with in-depth and up-to-date information, and to disseminate the results in academic circles and to the general population. She compiled statistics by gender and curriculum of university student enrollment in the province from 1999 to 2005. Her team also collected data on women professors and on the rates of women in the Canada Research Chairs program in order to document their low success rates. Although gender-disaggregated statistics were essential to the understanding of the progression of women in science and technology, Claire quickly realized that she would need to collaborate with various research groups in sociology and education in order to fully understand the problem. Together with the Centre interuniversitaire de recherche sur la science et la technologie (CIRST—the interuniversity center for research on science and technology), she participated in the qualitative analysis of school perseverance, curriculum changes, and dropout rates in science and technology studies at the college level. In another collaboration with the Comité Femmes en ingénierie of the Ordre des ingénieurs du Québec (OIQ – the Order of Engineers of Quebec), the Chair engaged in three studies confirming the difficulties of OIQ women members. Salary differences, work-family balance, and recognition of skills were among the key challenges reported by participants. Overall, the Chair's research led to a better knowledge of the process of selection – or more precisely the non-selection – of scientific careers by young people in general and young women in particular. These studies allowed her team to understand that young women choose their careers and persevere in different ways than men, depending on their background, the pressure of their peers and entourage, as well as various factors related to their school experience. This groundwork helped Claire determined which interventions would be the most effective in promoting diverse career choices to girls.[25]

[24] Caroline D'Amours, "'Une conquête inachevée': L'État québécois dans la promotion des femmes en science et génie, 1979–2003", paper presented at the *Canadian Science and Technology Historical Association Biennal Conference XVII*, November 19, 2011, Ottawa, Canada; Claire Deschênes, *Bilan final du second mandat de la Chaire CRSNG/Alcan pour les femmes en sciences et génie au Québec*, Québec, Université Laval, janvier 2007, p. 6; Claire Deschênes, Interview with Hannah Elizabeth Young, telephone interview, March 13, 2018, and Monique Frize, Claire Deschênes, Elizabeth Cannon, Mary Williams, and Marie Klawe, "A Unique National Project to Increase the Participation of Women in Science and Engineering" in Barbara Bogue, Priscilla Guthrie, Barbara Lazarus, and Steve Hadden, *Tackling the Engineering Resources Shortage: Creating New Paradigms for Developing and Retaining Women Engineers*, Bellingham, International Society for Optical Engineering, 1999, p. 88.

[25] Claire Deschênes, *Bilan final du second mandat de la Chaire CRSNG/Alcan pour les femmes en sciences et génie au Québec*, Québec, Université Laval, janvier 2007, p. 6 and 16–17.

Outils Pédagogiques
Utiles en Sciences
www.fsg.ulaval.ca/OPUS

As part of its program development, the Chair worked on the creation of the *Outils pédagogiques utiles en science* (OPUS), a web-based resource for high school teachers. Research showed that there were two main reasons explaining girls' lack of interest for physical sciences and engineering fields: their unfamiliarity with these career possibilities and their negative high school experience with these subjects. To answer these concerns, the OPUS website offered career guidance for young people, more specifically girls. The goal was to improve information that appeared to be deficient in content and dissemination, which, in turn, prevents young people from making informed career choices. Additionally, the website addressed students' criticism of science teaching by providing basic information, activities, and hands-on projects to help teachers in their difficult task. The website followed the high school physics curriculum and presented the teaching objectives with a science, technology, and society approach: the teaching objectives were linked to topics more closely related to young people's lives and interests, like breathalyzers for chemical dilutions, the physics of biking, or laser and myopia. High school teachers were involved to ensure the relevance and level of content.[26]

The OPUS project was a great success story. From its early design in 1999, through its launch in 2001, and to its conclusion in 2005, its more than 10,000 hours of work led to the development of original, high-quality scientific, and educational material, including more than 660 pages, 30 interactive exercises, 80 PDF files, and 830 images. OPUS not only made sciences accessible, fun, and exciting, but this project also reached its target audience all over the province. According to a 2004 survey, one third of the visitors were teenagers, of whom 78% were girls. Besides Quebec teachers consulting OPUS, visitors from French-speaking countries like France, Switzerland, and Belgium wandered through the website as well.[27]

[26] Claire Deschênes, *Bilan final du second mandat de la Chaire CRSNG/Alcan pour les femmes en sciences et génie au Québec*, Québec, Université Laval, janvier 2007, p. 10–11; Claire Deschênes, Interview with Hannah Elizabeth Young, telephone interview, March 13, 2018; Barbara Muir, NSERC, "The NSERC/Industry Chairs for Women in Science and Engineering: A National Program in Canadian Universities", paper presented to 12th International Conference of Women Engineers and Scientists, Ottawa, July 2002, accessed November 28, 2018, https://www.mun.ca/cwse/icwes_cwse5.pdf and Claire Deschênes, Martine Foisy, Judith Sévigny, and Anne-Marie Lemay, "Outils pédagogiques utiles en sciences (OPUS): Intéresser davantage les filles du secondaire aux sciences physiques", in Louise Lafortune and Claudie Solard (eds.), *Femmes et maths, sciences et technos*, Sainte-Foy, Presses de l'Université Laval, 2003, p. 127–146.

[27] Claire Deschênes, *Bilan final du second mandat de la Chaire CRSNG/Alcan pour les femmes en sciences et génie au Québec*, Québec, Université Laval, janvier 2007, p. 10–11.

Unfortunately, with the implementation of the Quebec educational reform in the mid-2000s, OPUS became obsolete almost overnight. Given that Claire did not have the resources to undertake the necessary work to adapt the website to the new curriculum at the time, the reform put an immediate end to a project that was dear to her heart.[28]

Claire also established *Les Midis-rencontres de la Chaire CRSNG/Alcan pour les femmes en sciences et génie*, a series of monthly brown-bag lunches offered to women graduate students of the three faculties of science at Université Laval. The Chairholder recognized that women graduate students needed to have discussions with role models to develop their own specific expertise related to research and its professional practice. The meeting format was simple. It started with the presentation of a chosen subject by a specialist, followed by a panel of women from different backgrounds talking about their personal experience, and then concluded by a period of lively discussions. Topics included components of the research job such as preparing for a career in research, developing a research program, finding a job after a Master's or doctorate degree, writing a thesis and scientific articles, and dealing with social and ethical issues of research dissemination in science and technology. More sensitive issues such as being a foreign woman and immigrant in STEM, reconciling family, studies and work, and exercising leadership were also discussed. According to testimony presented in Claire's final report, the response from the students seemed to have been very favourable. More than enthusiastic about these meetings, they even came forward with several suggestions for future topics.[29]

Claire additionally contributed to the creation of the *Association de la francophonie à propos des femmes en sciences, technologies, ingénierie et mathématiques* (AFFESTIM), a francophone non-profit association whose women directors came from five provinces of Canada (New Brunswick, Ontario, Nova Scotia, Prince Edward Island, and Quebec). Following a resolution taken at the Quebec roundtable during the ICWES 12 conference in July 2002, Louise Lafortune, professor in the Department of Educational Sciences at Université du Québec à Trois-Rivières and member of the Mouvement international pour les femmes et l'enseignement des mathématiques, and Marie Bernard, professor in the Department of Mechanical Engineering at École Polytechnique and Chairholder of the Marianne-Mareschal Chair, joined forces with Claire to launch AFFESTIM in October 2003. Its mandate was to bring together natural and legal persons, in the Francophonie, interested in promoting the participation, retention, and advancement of women in fields related to STEM. Keeping in mind the possibility that the NSERC/Alcan Chair activities

[28] Claire Deschênes, Interview with Hannah Elizabeth Young, telephone interview, March 13, 2018.

[29] Anne Roy, Claire Deschênes, Geneviève Boisclaire-Châteauvert, and Caroline Simard, "Un blogue d'information et de discussion pour inciter les jeunes femmes à s'orienter et persévér en STIM", in Anne Roy, Donatille Mujawamariya, and Louise Lafortune. *Des actions pédagogiques pour guider des filles et des femmes en STIM. Sciences, technos, ingénierie et maths.* Québec, Presses de l'Université du Québec, 2014, p. 136 and Claire Deschênes, *Bilan final du second mandat de la Chaire CRSNG/Alcan pour les femmes en sciences et génie au Québec*, Québec, Université Laval, janvier 2007, p. 14–15.

could cease in the future, this association also aimed to take over the knowledge, artifacts, and documentation center established during Claire's eight-year term. In only three years, AFFESTIM took a leadership role in the province, presenting a memoir to the Quebec Minister of Employment and Social Solidarity, publishing three electronic newsletters, organizing two symposia on women in STEM with the publication of the conference proceedings as part of the Association francophone pour le savoir (ACFAS) congress, and building representation throughout the province. In 2007, AFFESTIM was recognized among the major organizations deemed crucial to the promotion of scientific culture in the STEM-related fields supported by the Quebec Department of Economic Development, Innovation, and Export Trade. Although this association was small, and still is today, it brought together diverse francophone associations, exerted a strong leadership, and linked its members to the INWES network.[30]

AFFESTIM

In addition to her work with INWES and AFFESTIM, Claire also collaborated with other international organizations such as UNESCO, the Organisation for Economic Cooperation and Development (OECD), and the European Union. For example, Claire has been an advisory member of the Canadian Commission for UNESCO since the World Conference on Science of Budapest in 1999. As such, she was invited to participate in several events like the preparation of UNESCO's *International Report on Science, Technology, and Gender* in Marrakech, Morocco, in 2005. She also collaborated twice with the OECD Ad Hoc Working Group on the Steering and Funding of Research Institutions (SFRI) which has been addressing the issue of women in science specifically since 2005. Finally, Claire participated in the evaluation of the applications as an international expert for the European Union platform for women scientists. Her involvement in international organizations helped her develop a deeper understanding of the issue. She also acquired better knowledge of programs used by other countries to increase the participation of women in science and engineering, thus providing a greater understanding of what works and what doesn't. This fed her own research and development of her program in Quebec.[31]

---

[30] Claire Deschênes, *Bilan final du second mandat de la Chaire CRSNG/Alcan pour les femmes en sciences et génie au Québec*, Québec, Université Laval, janvier 2007, p. 13–14; Claire Deschênes, Interview with Hannah Elizabeth Young, telephone interview, March 13, 2018, and Claire Deschênes, "AFFESTIM: Son leadership", in Louise Lafortune, Claire Deschênes, Marie-Claude Williamson, and Pauline Provencher (eds.), *Le leadership des femmes en STIM: Sciences, technologies, ingéniérie et mathématiques*, Québec, Presses de l'Université du Québec, 2008, p. 147.

[31] Claire Deschênes, *Bilan final du second mandat de la Chaire CRSNG/Alcan pour les femmes en sciences et génie au Québec*, Québec, Université Laval, janvier 2007, p. 22 and Claire Deschênes, Interview with Hannah Elizabeth Young, telephone interview, March 13, 2018.

The Atlantic Chair: F. Mary Williams (1997–2002)

As a Senior Research Officer in the Institute for Marine Dynamics of the National Research Council of Canada, Florence-Mary Williams was the perfect candidate to be appointed as the NSERC/Petro-Canada Chair in WISE in the Atlantic Provinces and the Northwest Territories. An experienced mathematician, Mary traveled widely over the circumpolar seas to analyze the performance and safety of ocean engineering systems and determine the effects of ice on ships and offshore structures in the cold ocean environment. In addition to her research achievements, she also actively participated in many activities to encourage young women to take up careers in science and engineering. With her nomination as Chair in September 1997, she joined the Engineering and Applied Science faculty and the Department of Physics and Physical Oceanography at Memorial University in St. John's, Newfoundland.[32]

Mary's work as the Atlantic Chair was influenced by specific conditions in the Atlantic region. First, the presence of several strong organizations, like the Newfoundland and Labrador WISE chapter, the Women in Resource Development Committee (the WRDC – an organization whose goal was and still is to promote women in the oil and gas sector), and the Association of Nova Scotia Women for Education and Research in Science (ANSWERS – a sister organization to WISE), gave her the possibility of building strong partnerships right from the beginning. Partly the result of Monique Frize's work as National Chair at UNB, these partnerships would extend the Chair's reach in the Atlantic region. Second, Petro-Canada, the industry sponsor of the Chair, was actively engaged throughout Mary's term. Consequently, the Chair did some specific work related to women in the oil and gas sector. For example, the CCWEST conference held in St. John's in 2000 included a

[32] Monique Frize, Claire Deschênes, Elizabeth Cannon, Mary Williams, and Marie Klawe, "A Unique National Project to Increase the Participation of Women in Science and Engineering" in Barbara Bogue, Priscilla Guthrie, Barbara Lazarus, and Steve Hadden, *Tackling the Engineering Resources Shortage: Creating New Paradigms for Developing and Retaining Women Engineers*, Bellingham, International Society for Optical Engineering, 1999, p. 85 and Five New Chairs Press Release, p. 1, 1997, 10–123, Box: NSERC Chairs, Folder: News Release Federal, Monique Frize Fonds, University of Ottawa Archives and Special Collections.

panel focusing on women in this industry with representatives from a number of the large companies active in the Atlantic region. The post-event report provided a baseline standard for all companies in this sector, which led to other activities with the industry. Another major influence on the Chair's work was the requirement, since the signature of the Atlantic Accord in 1985, that any major natural resources sector projects taking place in Newfoundland and Labrador include a Workplace Diversity Plan and a Women's Employment Plan. This further steered the Chair to work with the oil and gas sector and, more specifically, with the WRDC. This requirement also engaged the Chair to maintain close contacts with Newfoundland and Labrador government departments: many senior civil servants were in fact part of the Chair's Advisory Board.[33]

An applied mathematician by training, Mary put her rational thinking immediately to work upon taking up her new position. Given the broad and ambitious objectives, the considerable previous activity by other agencies in the Atlantic region, and the scarce resources allocated to her Chair, she needed to plan how to best apply her efforts and resources. Without delay, Mary and her professional assistant, Carolyn Emerson, initiated an examination of participation rates in discipline and education level subgroups and gathered knowledge from the scientific literature about factors influencing these rates. After reviewing these data, she adopted a clear vision for her Chair: participation of women in science and engineering needed to reach at least 33% of any group or organization in the Atlantic region, especially at the decision-making levels. In order to put her vision to work, she devised a strategic plan which included three elements: leadership, excitement, and new traditions.[34]

In Mary's mind, the first ingredient necessary to change attitudes toward women in these fields was the presence of leaders for change. In order to do that, the Chair needed to support the identification, development, and exercise of leadership by women. Thus, Mary instigated two major initiatives in career development in collaboration with WISE Newfoundland and Labrador. The first was the creation of the *CWSE/WISE Becoming Leaders Workshops* on career topics designed for women early in their professional careers in science, engineering, and technology, and senior students in these fields. The workshop development began at the *CCWEST* conference held in St. John's in July 2000, where five of the 15 sessions addressed career

---

[33] Carolyn Emerson, Interview with Hannah Elizabeth Young, telephone interview, January 3, 2018; F.M. Williams and C.J. Emerson, *48–month progress report submitted to Natural Sciences and Engineering Research Council of Canada*, St. John's, Memorial University, March 2001, p. 9–14, and Stephanie Porter and the Diversity Network, *Using balance to build: Supporting Gender Diversity in Newfoundland and Labrador Construction Trades, 1990–2017*, St. John's, WRDC, 2018, p. 5, available at https://wrdc.ca/app/uploads/2018/03/Balance_To_Build.pdf

[34] Carolyn Emerson, Interview with Hannah Elizabeth Young, telephone interview, January 3, 2018; F.M. Williams, "The CWSE Strategy for the New Millenium", address presented to the *CCWEST Conference*, Vancouver, May 22, 1998, available at http://www.mun.ca/cwse/millenium.html and F.M. Williams and C.J. Emerson, *48-month progress report submitted to Natural Sciences and Engineering Research Council of Canada*, St. John's, Memorial University, March 2001, p. 3–4.

topics. The high attendance at the conference and the overwhelming positive feedback convinced Mary that career workshops really met women's needs. As a result, the Chair developed four different three–six-hour workshops: *Becoming Leaders*, *Negotiating for Success*, *Career Success*, and *Working for Success*. In addition to providing crucial information for women's early careers, the workshops enabled women to develop a network and gave them an opportunity to discuss and recognize common features of their experiences. Indeed, many women did not recognize that they were at a disadvantage because of their gender. Most of them had never been able to discuss their experiences with another woman because of the absence of a network.[35]

The original plan was to continue the development of the various modules through 2002, and then to package them for remote delivery in Mary's second term. However, her appointment as Director General for the National Research Council Institute for Marine Dynamics in 2002 put a pause in the development of the series. Fortunately, Mary allowed the Canadian Centre for Women in Science, Engineering, Trades, and Technology (WinSETT Centre) to take over these workshops. Since then, the WinSETT Centre has developed a series of six workshops that are being offered across the country and address essential topics in career development in addition to one module intended for employers. As of November 2018, the WinSETT Centre had delivered over 130 workshops and 30 special topics that had received positive evaluation from more than 4,000 participants.[36]

The second major initiative of Mary's leadership strategy was the publication, in partnership with WISE, of *Becoming Leaders: A Handbook for Women in Science and Engineering and Their Mentors* in June 2002. The Handbook was first based on the information and feedback obtained in the workshop series. The book provided accessible and practical information to professional women and students in science and engineering, and managers interested in the development of human resources in their organization. It covered topics identified as women's chief concerns: work-life balance as well as fair evaluation and advancement. Supported by knowledge drawn from research in the humanities and social sciences fields, the handbook presented the information in a concise and straightforward manner. Approximately 800 copies sold in the first four months after its publication. An American edition of the handbook was published in 2008 by the Society of Women Engineers, the American Society of Civil Engineers Press, and the American Society of Mechanical Engineers

[35] F.M. Williams and C.J. Emerson, *48-month progress report submitted to Natural Sciences and Engineering Research Council of Canada*, St. John's, Memorial University, March 2001, p. 6–7; F.M. Williams and C.J. Emerson, *Atlantic Region Chair for Women in Science and Engineering: 2001–2002 Report*, St. John's, Memorial University, May 2002, p. 1–3 and Carolyn Emerson, Interview with Hannah Elizabeth Young, telephone interview, January 3, 2018.

[36] F.M. Williams and C.J. Emerson, *Atlantic Region Chair for Women in Science and Engineering: 2001–2002 Report*, St. John's, Memorial University, May 2002, p. 1–3; Carolyn Emerson, Interview with Hannah Elizabeth Young, telephone interview, January 3, 2018, and WinSETT Centre, "Leadership Program", accessed January 27, 2019, http://www.winsett.ca/Programs/leadership-program

Press. Carolyn Emerson, Mary's former professional assistant as Chairholder, worked on the second American edition which appeared in 2019.[37]

The second component of Mary's strategy was excitement: to motivate women to enter and remain in science and engineering. The common approach to enhancing public perception of fields of study at the time was to talk about the challenge, income, or sense of self-worth provided by these fields. But Mary wanted instead to express the excitement and satisfaction enjoyed by scientists and engineers because this was the actual motivation behind their work. To achieve this goal, she made presentations to high school students, both boys and girls, where she emphasized the fun of discovering and learning in science and the satisfaction of making a contribution to one's community. She also carried the same message to career events with Girl Guides and leadership training events with Guide leaders. The Chair additionally worked in partnership with WISE to found the WISE Student Summer Employment Program, which provided female high school students with real work experience in science and engineering. With approximately 50% of program graduates entering postsecondary science and engineering studies, the program was highly successful.[38]

The third ingredient of her plan was new traditions. This strategy recognized that "many of the factors which overtly or subtly deter women from science

[37] F.M. Williams and C.J. Emerson, *48–month progress report submitted to Natural Sciences and Engineering Research Council of Canada*, St. John's, Memorial University, March 2001, p. 6; F.M. Williams and C.J. Emerson, *Atlantic Region Chair for Women in Science and Engineering: 2001–2002 Report*, St. John's, Memorial University, May 2002, p. 1–2; and Carolyn Emerson, Interview with Hannah Elizabeth Young, telephone interview, January 3, 2018.

[38] F.M. Williams and C.J. Emerson, *48–month progress report submitted to Natural Sciences and Engineering Research Council of Canada*, St. John's, Memorial University, March 2001, p. 7.

and engineering careers are not explicit policy or statement, but rather they are deep and systemic factors in institutions" where women worked.[39] For this reason, the Chair's objectives were to identify these traditions and communicate to institutional leaders the value of diversity in order to give them the rationale to implement change. As part of her research, Mary analyzed what were the axioms that formed the basis of science and engineering thinking, in particular with respect to communication and evaluation of merit. A second thread of investigation reviewed the understanding of the diversity of styles with which people manage scientific and technical information and the possibility of gender patterns. Mary was also interested in the analysis of the flow of women into science and engineering fields. This last study argued that a minority ratio of women tends to further diminish as the cohort advances through a career because institutions of higher education and their academic traditions are well sheltered from change influences. Finally, the Chair examined the ways in which scientists think and considered whether thinking styles are a factor in determining who participates successfully in science. The infographic forecast drawn from this research also provided a strong argument in favour of diversity in science.[40]

The *New Frontiers, New Traditions CCWEST* Conference held in 2000 was the final component of Mary's new traditions strategy. Thirty-three papers presented research results and case studies. Collectively, the papers represented up-to-date understanding of the complex gender dynamics of people in science and engineering work and study environments. Some of the papers presented were the first Canadian benchmarks for issues that had been explored in other countries. Other papers provided entirely new perspectives. In addition to gathering 248 women and men from Canada, the United States, and Sweden, the conference results were strongly positive. Information about the promotion and retention of women in science and engineering, particularly in the oil and gas industry and the academic sector, was also consolidated through the presentations and discussions. Women in the early stages of their career also had several opportunities to receive encouragement and practical tips for their professional paths. The participation and sponsorship from various organizations also sent a strong message to both conference participants and other organizations in their sector. Finally, the conference generated long-terms results. As we have seen, the conference was the starting point for both the

[39] F.M. Williams and C.J. Emerson, *48-month progress report submitted to Natural Sciences and Engineering Research Council of Canada*, St. John's, Memorial University, March 2001, p. 4.

[40] Carolyn Emerson, Interview with Hannah Elizabeth Young, telephone interview, January 3, 2018; F.M. Williams, "The CWSE Strategy for the New Millenium", address presented to the *CCWEST Conference*, Vancouver, May 22, 1998, available at http://www.mun.ca/cwse/millenium.html and F.M. Williams and C.J. Emerson, *48-month progress report submitted to Natural Sciences and Engineering Research Council of Canada*, St. John's, Memorial University, March 2001, p. 3–4.

handbook and the workshops, and both projects are still very active over 20 years later.[41] As Carolyn Emerson noted, they are Mary's "greatest legacy."[42]

NEW FRONTIERS
New Traditions

The West Coast Chair: Maria Klawe (1997–2003)

Prior to her appointment as the NSERC/IBM Chair for Women in Science and Engineering in British Columbia and the Yukon, Maria Klawe was already an accomplished mathematician and computer scientist as well as an experienced academic administrator. She had spent eight years with IBM Research in California and two years as a faculty member in the Department of Computer Science at the University of Toronto. In 1988, she joined the Department of Computer Science at the University of British Columbia and served as its first female head from 1988 to 1995. At the time of her nomination as Chair in 1997, she was serving as Vice President of Student and Academic Services, since 1995. In addition to her administrative functions, Maria made significant research contributions in several areas of mathematics and computer science, including functional analysis, discrete mathematics, theoretical computer science, and interactive multimedia for mathematics education. She has long been passionate about increasing women's participation in computer science. As early as 1991, Maria and a fellow Computing Research Association (CRA) board member, Nancy Leveson, established CRA's Committee on the Status of Women in Computing Research "to take positive action to increase the number of women participating in Computer Science and Engineering (CSE) research and education at all levels."[43] She knew that computer science and technology were great career opportunities, offering good income, work-life balance, and possibilities to travel. They also offered a chance to make significant contributions to important societal problems, something women tend to be adamant about. Monique's work as National Chair attracted her to be involved in women's issues.

---

[41] F.M. Williams and C.J. Emerson, *48–month progress report submitted to Natural Sciences and Engineering Research Council of Canada*, St. John's, Memorial University, March 2001, p. 8–9.

[42] Carolyn Emerson, Interview with Hannah Elizabeth Young, telephone interview, January 3, 2018.

[43] Maria Klawe, Interview with Hannah Elizabeth Young, telephone interview, September 4, 2018, and Computing Research Association, "The history of CRA-W", accessed January 27, 2019, https://cra.org/cra-w/history/#toggle-id-3

So, when the opportunity to devote her time to this particular problem arose, with the creation of the five regional Chairs, she jumped right into it.[44]

Being a computer scientist herself, Maria decided right away to focus on increasing women's participation in information technology and computer science. She had organized her Chair activities into a five-year project known as the Supporting Women in InFormation Technology (SWIFT). The presence of the IBM Canada Pacific Development Centre in Vancouver and the fact that IBM was also the Chair's industrial partner probably pushed Maria further in this direction. The shortage of personnel faced by organizations in the information technology sector was most likely another factor influencing her decision. Indeed, the industry was in an increasingly intense need for more highly skilled people in the late 1990s and early 2000s. Yet, despite the existence of numerous well-paying job opportunities, few women enrolled in information technology and computer science programs. While women's participation in every area of science and engineering was increasing over the previous decade, the percentage and total number of Bachelor's degrees in computer science awarded to women was in fact decreasing almost every year over the same period.[45]

The SWIFT project was engaged in "researching, creating and disseminating resources and strategies to encourage sustained interest and achievement by girls

[44] Maria Klawe, "How Can We Encourage More Women to Study Computer Science?", CRA-W website, accessed January 27, 2019, https://cra.org/cra-w/629/ and Maria Klawe, Interview with Hannah Elizabeth Young, telephone interview, September 4, 2018.

[45] Rhian Davies, Maria Klawe, Michele Ng, Crystal Nyhus, and H. Sullivan, "Gender Issues in Computer Science Education" in *Proceedings of the National Institute of Science Education Forum*, Detroit, United States, 2000, p. 1, available at http://archive.wceruw.org/nise/News_Activities/Forums/Klawepaper.htm; Maria Klawe, Ian Cavers, Fred Popowich, and Grace Chen, "ARC: A Computer Science Post-Baccalaureate Diploma Program that Appeals to Women", in Ellen Balka and Richard Smith, *Women, Work and Computerization: Charting a Course to the Future*, Boston, Kluwer Academic Publishers, 2000, p. 94–96 and Monique Frize, Claire Deschênes, Elizabeth Cannon, Mary Williams, and Marie Klawe, "A Unique National Project to Increase the Participation of Women in Science and Engineering" in Barbara Bogue, Priscilla Guthrie, Barbara Lazarus, and Steve Hadden, *Tackling the Engineering Resources Shortage: Creating New Paradigms for Developing and Retaining Women Engineers*, Bellingham, International Society for Optical Engineering, 1999, p. 86.

and women of all ages in areas related to information technology."[46] The program was divided into four core initiatives. The first initiative was a survey to increase the understanding of high school students' perceptions of computer-related subjects and careers, as well as factors they perceive as influencing their career choices. There were two reasons to conduct this survey: first, to establish where SWIFT initiatives should focus when trying to increase the number of female students choosing to study computer science and second, to form a baseline with which to compare the future effects of SWIFT initiatives. Thus, in the fall term of 1998, 7,411 surveys were completed by students enrolled in Grades 8, 10, and 12 in all 18 schools of the Vancouver School District. The results of the survey indicated that female students were less interested and perceived that they had less ability in computer science than did the male students. Additionally, the study revealed that women spent less time playing computer games, programming, and surfing the Internet than their male counterparts, suggesting that spending less time using computers could be correlated with lower perceived ability in computer science and engineering. Finally, students' inaccurate perceptions of what skills and characteristics were necessary for success in computer-related careers seemed to be another factor explaining girls' low interest and perceived ability. The fact that students did not know that recruiters from the information technology industry were looking first and foremost for workers with strong communication and teamwork skills, two areas in which women were viewed as having particular strength in North America, could explain why women were not interested in computer-related careers in the first place.[47]

Virtual Family was the second initiative initiated by Maria and members of the SWIFT team. Studies had shown that female students' lack of programming experience prior to entering university was a contributing factor to their low participation and high attrition rates in computer science courses when compared to their male counterparts. With *Virtual Family*, the Chair attempted to develop new material and approaches to make learning programming more fun, accessible, and attractive to girls with no programming experience. This one–two-hour workshop targeting teenage girls mixed Java programming with components of electronic games that research had shown to appeal to girls, namely story lines, characters, and relationships. The *Virtual Family* storyline was based on a family and their interactions. It included four characters: a mother, a father, a daughter, and a son in addition to family pets in the case where students chose to include them. In a user-friendly Java programming environment, girls and boys were able to make the characters perform

[46]V. Chan, K. Stafford, Maria Klawe, and Grace Chen, "Gender Differences in Vancouver Secondary Students: Interests Related to Information Technology Careers", in Ellen Balka and Richard Smith, *Women, Work and Computerization: Charting a Course to the Future*, Boston, Kluwer Academic Publishers, 2000, p. 58.

[47]V. Chan, K. Stafford, Maria Klawe, and Grace Chen, "Gender Differences in Vancouver Secondary Students: Interests Related to Information Technology Careers", in Ellen Balka and Richard Smith, *Women, Work and Computerization: Charting a Course to the Future*, Boston, Kluwer Academic Publishers, 2000, p. 60–68.

various actions and have other characters react to these actions. In contrast to the traditional and monotonous "Hello world" introduction to programming, *Virtual Family* made the game more appealing to girls by using graphics, sound, animation, and color. This program received favourable feedback and showed multiple beneficial impacts. It increased the awareness of computer science as a possible career option, raised the level of programming experience students had before entering university, and decreased students' computer anxiety.[48]

As a third major initiative, Maria and the SWIFT team engaged in a variety of outreach activities aiming at changing attitudes and images about computer science. The goal was also to include the extended community in raising awareness of the issues and expand the base for brainstorming effective approaches. To achieve this, the Chair first supported visits by volunteer scientists and engineers to K-12 classrooms through the Scientists and Innovators in the Schools and *Pathmakers* programs that were already in place for the past decades. Maria and her team also organized information technology workshops for female students. These included activities like Ms. Infinity, a one-day hands-on conference for young women in Grades 9 and 10, and the IBM's Women in Technology workshops, a half-day discussion group held by female IBM staff and SWIFT volunteers. Finally, Maria gave several talks and met with faculty and students at Simon Fraser University and the University of British Columbia in addition to giving public lectures in Victoria and Prince George and at several conferences and workshops.[49]

The final and also one of the main efforts made by Maria to make university computer science programs more accessible to women was the *Alternate Routes to Computing* (ARC) program. ARC was a two-year cooperative education diploma program composed of four four-month terms of university-level courses, interspersed with two four-month terms of paid industry experience. The creation of an alternative computing program as a SWIFT initiative was first suggested in 1997 by James Lau, the Director of the IBM Canada Pacific Development Centre. His goal

---

[48] Rhian Davies, Maria Klawe, Michele Ng, Crystal Nyhus, and H. Sullivan, "Gender Issues in Computer Science Education" in *Proceedings of the National Institute of Science Education Forum*, Detroit, United States, 2000, p. 15–16, available at http://archive.wceruw.org/nise/News_Activities/Forums/Klawepaper.htm; Maria Klawe, "Engaging Female High School Students in Computer Programming via Virtual Family", accessed January 20, 2019, http://www.mun.ca/cwse/VFworkshop.pdf and Barbara Muir, NSERC, "The NSERC/Industry Chairs for Women in Science and Engineering: A National Program in Canadian Universities", paper presented to 12th International Conference of Women Engineers and Scientists, Ottawa, July 2002, accessed November 28, 2018, https://www.mun.ca/cwse/icwes_cwse5.pdf

[49] Monique Frize, Claire Deschênes, Elizabeth Cannon, Mary Williams, and Marie Klawe, "A Unique National Project to Increase the Participation of Women in Science and Engineering" in Barbara Bogue, Priscilla Guthrie, Barbara Lazarus, and Steve Hadden, *Tackling the Engineering Resources Shortage: Creating New Paradigms for Developing and Retaining Women Engineers*, Bellingham, International Society for Optical Engineering, 1999, p. 87 and Rhian Davies, Maria Klawe, Michele Ng, Crystal Nyhus, and H. Sullivan, "Gender Issues in Computer Science Education" in *Proceedings of the National Institute of Science Education Forum*, Detroit, United States, 2000, p. 15–16, available at http://archive.wceruw.org/nise/News_Activities/Forums/Klawepaper.htm

was to increase the number of information technology workers combining a significant level of knowledge and expertise outside this field, with technical knowledge and expertise to contribute to software application development in domains like biology, languages, and social work. The University of British Columbia and Simon Fraser University, together with the provincial government, joined forces in December 1997 to develop this fast-track computer science diploma program. Despite many difficulties encountered by the first cohort of students and organizers in 1998, students remaining in the program performed well academically and were well-regarded by their employers. The ARC program also met Maria's goal to increase women's participation in information technology, as over 70% of this cohort was composed of women students. Due to the demand and success of the ARC project in 1998, another cohort of students was accepted to the program in 2000. Since then, the program has been successful and it still exists today, although in a slightly different form, in both the University of British Columbia and Simon Fraser University.[50]

The Prairie Chair: Elizabeth Cannon (1997–2004)

When Elizabeth Cannon began her mandate as the Petro-Canada/NSERC Chair for Women in Science and Engineering for the three Canadian Prairie provinces (Alberta, Saskatchewan, and Manitoba), she had been a professor in the Department of Geomatics Engineering at the University of Calgary for six years. She was involved in numerous industry-sponsored research projects related to satellite-based radio navigation using the global positioning system (GPS) as well as technology transfer activities. The Montreal Massacre was a major turning point in her life. Up until then, she was mostly focused on her own work and was not really thinking a lot about the barriers that women in science and engineering were facing in 1989. This tragedy led her to reflect on her personal contribution to the community and to take action to support other women in her field.[51]

---

[50] Rhian Davies, Maria Klawe, Michele Ng, Crystal Nyhus, and H. Sullivan, "Gender Issues in Computer Science Education" in *Proceedings of the National Institute of Science Education Forum*, Detroit, United States, 2000, p. 1, available at http://archive.wceruw.org/nise/News_Activities/Forums/Klawepaper.htm; Maria Klawe, Ian Cavers, Fred Popowich, and Grace Chen, "ARC: A Computer Science Post-Baccalaureate Diploma Program that Appeals to Women", in Ellen Balka and Richard Smith, *Women, Work and Computerization: Charting a Course to the Future*, Boston, Kluwer Academic Publishers, 2000, p. 94–96; UBC website, "BCS Program (Second Degree)", accessed January 20, 2019, https://www.cs.ubc.ca/students/undergrad/programs/second-degree and Simon Fraser University website, Second Degree in Computing Science", accessed February 9, 2019, http://www.sfu.ca/computing/current-students/undergraduate-students/programs/second-degree-in-computing-science.html

[51] Monique Frize, Claire Deschênes, Elizabeth Cannon, Mary Williams, and Marie Klawe, "A Unique National Project to Increase the Participation of Women in Science and Engineering" in Barbara Bogue, Priscilla Guthrie, Barbara Lazarus, and Steve Hadden, *Tackling the Engineering*

When she joined the engineering school in 1991, she made sure to be involved in the university WISE chapter. Her role was crucial since she was only the second woman faculty member at the time. Her program as Chair bore the mark of her own identity. As an engineer at the University of Calgary, and through her connections with the Association of Professional Engineers and Geoscientists of Alberta (APEGA), it was easier for Elizabeth to have an impact on the engineering field. She focused primarily on her own local community. The strong commitment and encouragement from the University of Calgary and the Chair's sponsors helped Elizabeth achieve many of the goals she initially set.[52]

The activities of the Prairie Chair were divided into three general areas: public awareness, data acquisition and research, and program development. In terms of public awareness, Elizabeth gave more than 200 presentations to promote engineering and science to students at high school and university levels (and their teachers), community groups, as well as industry and political leaders. Her presentations included up-to-date statistical information on the participation and performance of women in high school science and mathematics courses, the enrollment of women in university-based programs, the participation, promotion, and retention of women in industry, and results from her own research. The Chair was also able to raise public awareness through interviews given to the media resulting in articles in newspapers and magazines, as well as radio appearances. She developed a brochure to provide information about the Chair and its goals to teachers, counselors, community and professional groups, university administrators, and industry

---

Resources Shortage: Creating New Paradigms for Developing and Retaining Women Engineers, Bellingham, International Society for Optical Engineering, 1999, p. 85 and Ontario Chair for Women in Science and Engineering, "Our Archiving Conversation Continues with Dr. Elizabeth Cannon", accessed January 20, 2019, http://scieng-women-ontario.ca/en/2014/10/our-archiving-conversation-continues-with-dr-elizabeth-cannon-a-leaders-role-in-building-a-legacy-is-to-leave-behind-his-or-her-learnings-for-others-to-find-them-critique-them-and/

[52] Elizabeth Cannon, *Progress Report #2: NSERC/Petro-Canada Chair for Women in Science and Engineering (Prairies Region)*, août 2001, p. 14 and 23; Anne Millar and Mary Wells, "Engineering taught Elizabeth Cannon how to be a leader", *CIM Magazine*, October 30, 2017, accessed January 20, 2019, http://magazine.cim.org/en/voices/elizabeth-cannon/ and Elizabeth Cannon, Interview with Hannah Elizabeth Young, telephone interview, May 3, 2018.

personnel. A final means of creating awareness about the Chair and its activities was through the creation of a website. It also included statistical information, links to other relevant sites, and a bibliographic database on women in science and engineering research.[53]

With respect to data collection and research, Elizabeth directed her attention toward two broad spheres. The first major research project studied the perceptions and decision-making processes around science and mathematics for boys and girls in high school programs. It analyzed both qualitative and quantitative data drawn from a survey of 1,430 Grade 7 and 10 boys and girls from the Calgary region, in addition to 200 in-depth interviews of average and high-average ability students in Grade 7, and ten in-depth interviews of students' parents, as well as a survey completed by teachers and school counselors. The research revealed that female students, like their male counterparts, showed interest in challenging careers and were also open to pursuing high levels of education to enter these careers. Yet, unlike male students, they took into account the ability of the career to allow them to contribute to the world and exercise creativity when looking at career choices.[54] Elizabeth's second major research project focused on determining the key factors influencing both women and men to select engineering as a major in their university undergraduate program. Using data collected from a survey administered to 1,066 undergraduate engineering students at the University of Calgary, the results indicated that female students showed less interest and perceived ability in physics at both the high school and university levels although their academic performance was similar to their male counterparts. However, in the case of mathematics, women outperformed men and had significantly greater interest and perceived ability. The quality and impact of this research project was recognized by the US-based Women in Engineering and Programs Advocates Network (WEPAN) by presenting Elizabeth and her two collaborators with the Betty Vetter Award which acknowledged notable achievement in research related to women in engineering.[55]

During her term as Chair, Elizabeth developed several new programs and gave support to programs already in existence. Among the newly developed programs based on her own research data, two stand out. Created in 2001 in collaboration with the University of Calgary's Schulich School of Engineering, the University of Alberta, and the Alberta Women's Science Network, *SCIbermentor* aimed to inspire girls to pursue STEM careers. The program matched Alberta girls aged 11 to 18 on a one-on-one basis to women science and engineering students and professionals who served as role models and support. The program, renamed *Cybermentor*, still builds relationships between girls and mentors through a free digital platform as well as hands-on activities. As Qiao Sun, Associate Dean (diversity and equity) of

[53] Elizabeth Cannon, *Progress Report #2: NSERC/Petro-Canada Chair for Women in Science and Engineering (Prairies Region)*, Calgary, University of Calgary, août 2001, p. 5–6.

[54] Elizabeth Cannon, *Progress Report #2: NSERC/Petro-Canada Chair for Women in Science and Engineering (Prairies Region)*, Calgary, University of Calgary, août 2001, p. 7–8 et 23.

[55] Elizabeth Cannon, *Progress Report #2: NSERC/Petro-Canada Chair for Women in Science and Engineering (Prairies Region)*, Calgary, University of Calgary, août 2001, p. 9–10 et 23.

the Schulich School of Engineering observed in 2017, such "personal connections are proven to be more effective in helping girls bridge their passions to engineering fields, build confidence, and ultimately lead them to a rewarding career path."[56] Since its creation, *Cybermentor* has reached over 3,000 young women from 70 communities across Alberta. In 2017, the program received the WEPAN Women in Engineering Initiative Award, which recognizes an outstanding project or initiative that serves as a model for other institutions in efforts to increase the recruitment and retention of girls and women in engineering.[57]

CYBERMENTOR

A second project developed by the Prairie Chair was the *Explore IT* conferences for Grade 9 girls, an event focused on introducing girls who are about to enter high school to career opportunities in these fields. It targeted girls at a critical moment when they begin to make their career decisions and plan for the future. A collaboration between the academic sector (the University of Calgary, the Southern Alberta Institute of Technology, and Mount Royal College), industry (Nortel, IBM), the government (Alberta Innovation and Science), and local school boards, Explore IT first brought together 350 Grade 9 girls in 1999 for a day of hands-on sessions and talks from keynote speakers. The first two conferences received extensive positive feedback from participants, the increase in attendance from year one to year two being a testimony of its success. A vast majority of the participants reported seeing computers and careers with computers in a more positive light. As of 2018, *Explore STEM* (as it was later known) had involved more than 8,000 students. Indeed, more than three quarters of the 2017 participants indicated they would consider pursuing a career in STEM. However, the program ended in 2019.[58]

---

[56] Alison Barrett, "Online Mentoring Program Honoured for Inspiring Girls to Pursue STEM Careers", June 29, 2017 available at https://www.ucalgary.ca/utoday/issue/2017-06-29/online-mentoring-program-honoured-inspiring-girls-pursue-stem-careers

[57] Elizabeth Cannon, *Progress Report #2: NSERC/Petro-Canada Chair for Women in Science and Engineering (Prairies Region)*, Calgary, University of Calgary, août 2001, p. 10–11 and Alison Barrett, "Online Mentoring Program Honoured for Inspiring Girls to Pursue STEM Careers", June 29, 2017 available at https://www.ucalgary.ca/utoday/issue/2017-06-29/online-mentoring-program-honoured-inspiring-girls-pursue-stem-careers

[58] Elizabeth Cannon, *Progress Report #2: NSERC/Petro-Canada Chair for Women in Science and Engineering (Prairies Region)*, Calgary, University of Calgary, août 2001, p. 12 and Peter Glenn, "Grade 9 girls Explore STEM", May 18, 2018 available at https://www.mtroyal.ca/AboutMountRoyal/MediaRoom/Newsroom/grade-9-girls-explore-stem.htm

Central to the longevity of these two initiatives was Elizabeth's ability to mobilize leaders from various sectors. She built collaborations and partnerships within the university and with the community in terms of developing the programs and resourcing them. Not only did she convince a diverse group of individuals and organizations of the relevance of the issue of participation of women in science and engineering, but she was also able to persuade them to provide considerable funding to her initiatives. Indeed, she first secured external funding from industry and government to cover the $30,000 of expenditures required to run the initial Explore STEM conference expenditures in 1999. The program was later sponsored by NSERC and industry actors like Cisco, Xerox, Encana, and Brand Alliance. She was also able to raise money for the inception of the *Cybermentor* in 2001. A $240,000 project, *Cybermentor*, was created thanks again to the external funding from industry and government actors such as Alberta Innovation and Science through their Science Awareness and Promotion grant program and PanCanadian Petroleum. This program still exists today with grants from Alberta Economic Development and Trade Science Awareness Program, NSERC PromoScience, and ConocoPhillips.[59]

Struggles Faced by the Five Chairs

As we have seen, the establishment of the five regional Chairs for Women in Science and Engineering gave a unique opportunity for fresh analysis of the concepts and processes which lead individuals to successful participation in these fields. Feeding off their own research, each Chairholder was able to share their data with university, government, and industry leaders in addition to developing various programs targeting girls and women. Furthermore, they dedicated a fair amount of their time to raising public awareness about this issue. Yet, the advantages derived from their unique positions and perspectives were balanced by the challenges specific to their role. The broad mandate of the Chair was the first of these challenges: the five

[59] Elizabeth Cannon, *Progress Report #2: NSERC/Petro-Canada Chair for Women in Science and Engineering (Prairies Region)*, Calgary, University of Calgary, août 2001, p. 10–12; Explore STEM, Meet the Explore STEM 2018 Sponsors", available at http://www.explore-it.ca/sponsors.html; Alison Barrett, "Online Mentoring Program Honoured for Inspiring Girls to Pursue STEM Careers", June 29, 2017 available at https://www.ucalgary.ca/utoday/issue/2017-06-29/online-mentoring-program-honoured-inspiring-girls-pursue-stem-careers and Elizabeth Cannon, Interview with Hannah Elizabeth Young, telephone interview, May 3, 2018.

women had only a five-year term and limited resources to tackle such a large and complex issue. Indeed, the program's objectives brought them to consider the societal, cultural, and educational aspects of women's participation in science and engineering. However, identifying the factors behind women's low participation in these fields and changing the educational, academic, and workplace culture was a difficult and long-term process that required patience. It must often have felt like a daunting prospect for scientists and engineers like themselves who were used to progressing through research and results to satisfactory solutions in a timely and planned manner.[60]

The five women also struggled with the feeling of isolation and marginalization experienced as a result of their involvement as Chairs, a phenomenon well known among people who work on behalf of the status of women. At times, the backlash against the promotion of women in science, engineering, and technology broke out in the public sphere as we have seen with the introduction of NSERC's University Faculty Awards in 1997. But, most of the time, it was experienced on a more personal level. Their engagement to actively promote women in science and engineering had occasionally been met with indifference in the academic sector, an environment often resistant to change. Frequently, their male colleagues reacted negatively because they thought that women were not interested in science or not good at it because of their "natural" attributes. To teach the value of diversity in the academic and workplace sector was a long-term process. Besides, in a male-oriented discipline, their gender research and promotion activities were often considered less prestigious. In addition to these misunderstandings, some Chairholders faced outright hostility. For example, Monique Frize had trouble getting an NSERC grant for several years due to personal attacks made by reviewers on the electrical engineering committee. In the end, maintaining respect and recognition from the mainstream actors in their respective disciplines was a time- and energy-consuming challenge for every Chairholder.[61]

---

[60] F.M. Williams and C.J. Emerson, *48-month progress report submitted to Natural Sciences and Engineering Research Council of Canada*, St. John's, Memorial University, March 2001, p. 4; Barbara Muir, NSERC, "The NSERC/Industry Chairs for Women in Science and Engineering: A National Program in Canadian Universities", paper presented to 12th International Conference of Women Engineers and Scientists, Ottawa, July 2002, accessed November 28, 2018, https://www.mun.ca/cwse/icwes_cwse5.pdf; Carolyn Emerson, Interview with Hannah Elizabeth Young, telephone interview, January 3, 2018, and Elizabeth Cannon, Interview with Hannah Elizabeth Young, telephone interview, May 3, 2018.

[61] Barbara Muir, NSERC, "The NSERC/Industry Chairs for Women in Science and Engineering: A National Program in Canadian Universities", paper presented to 12th International Conference of Women Engineers and Scientists, Ottawa, July 2002, accessed November 28, 2018, https://www.mun.ca/cwse/icwes_cwse5.pdf; Monique Frize, *The Bold and the Brave: A History of Women in Science and Engineering*, Ottawa, University of Ottawa Press, 2009, p. 182; Monique Frize, Interview with Hannah Elizabeth Young, telephone interview, May 8, 2018; Maria Klawe, Interview with Hannah Elizabeth Young, telephone interview, September 4, 2018; Carolyn Emerson, Interview with Hannah Elizabeth Young, telephone interview, January 3, 2018; and Monique Frize, *NSERC/Nortel Women in Engineering Chair – 48 Month Report*, September 2001,

The third main challenge faced by the five Chairholders was the difficulty of creating a strong impact outside of their local community. Meeting the regional mandate was especially difficult for the Atlantic, Prairies, and West Coast Chairs since their organizations covered more than one province. For example, the Prairie Chair spans three jurisdictions, with three separate provincial governments, three different education structures, and three postsecondary systems. This was in direct contrast to the Ontario and Quebec Chairs, where there tended to be more consistency in approach and existing networks. In order to maximize the regional influence of the Chair and create an impact in areas where there was no constant presence, the Chairs covering multiple provinces needed to develop a network. Of course, they would often travel to other provinces, but they could not always be on the road. One solution to the problem of distance was the establishment of an advisory board. In the case of the Prairie Chair, the Advisory Council consisted of 22 representatives from all the major universities as well as members of the industry and education sectors. For the Atlantic Chair, the Advisory Board included representatives from secondary schools, universities, provincial government, resource development, information technology, science promotion, finance, and the industry sponsor. Both these structures seemed to have been an effective means of communicating information across the region while providing regional input to the Chair.[62]

If the regional mandate was especially difficult to fulfill for these three Chairs, it did not mean that the Ontario and Quebec regions were without challenges. Like their sister organizations elsewhere in Canada, the central Canada Chairs also needed to reach small pockets of rural populations spread around their province. With only limited funds and personnel, the Chairholders had to be creative. For this reason, most of the Chairholders were adamant to meet leaders in universities such as presidents or deans of science and engineering in their district to explain their activities, the problem, and the results of their research. In the case of the Quebec Chair, the establishment of the OPUS website was another way of addressing this problem. Claire Deschênes' additional solution was the creation of regional representatives from three francophone regions: Dr. Ève Langelier – Université de Sherbrooke, Dr. Duygu Kocaefe – Université de Chicoutimi and Dr. Anne-Marie Laroche – Université de Moncton in New Brunswick. In addition to increasing the network of individuals working on the issue in the francophone world, these representatives also developed their own activities. Thus, the presence of these three regional

p. 14, 10–123, Box: NSERC Chairs, Folder: *NSERC/Nortel Women in Engineering Chair – 48 Month Report*, Monique Frize Fonds, University of Ottawa Archives and Special Collections.

[62] Carolyn Emerson, Interview with Hannah Elizabeth Young, telephone interview, January 3, 2018; Elizabeth Cannon, Interview with Hannah Elizabeth Young, telephone interview, May 3, 2018; Elizabeth Cannon, *Progress Report #2: NSERC/Petro-Canada Chair for Women in Science and Engineering (Prairies Region)*, Calgary, University of Calgary, août 2001, p. 3 and F.M. Williams and C.J. Emerson, *48-month progress report submitted to Natural Sciences and Engineering Research Council of Canada*, St. John's, Memorial University, March 2001, p. 2.

representatives allowed the Chair to multiply her actions. A final solution to the problem of the regional mandate was to build collaboration or partnerships. This was the reasoning behind the financial support of the *Let's Talk Science* initiative by the Atlantic Chair. The same reasoning was behind the interest of the Prairie Chair with *Cybermentor* and *Explore STEM*. In both cases, it was a matter of building partnerships to leverage their energy and the small resources available and, ultimately, make a difference.[63]

***

In the end, the five Chairholders struggled above all to maintain momentum in scientific research and teaching while fulfilling their Chair mandate. In contrast to other NSERC-supported Chairs that were focused on the researcher's primary expertise, the CWSE program required Chairholders to commit approximately half their time to a second area of interest. However, the very active responsibilities of the Chair, with its leadership and promotion expectations and the high volume of public and individual communications, consumed more than 50% of their time for most of the Chairholders. This situation generated tensions between their work as Chairs and their regular research and teaching activities in their respective disciplines. In a very competitive environment like universities, the pressure must have been difficult to bear. For example, Claire Deschênes' scientific production in her primary area of expertise dropped so much in her final years as Chairholder that she did not bother to renew her NSERC Discovery Grant that was about to expire in April 2005. Yet, Claire and her fellow Chairholders committed a substantial portion of their career to making a contribution and building a reputation, as serious scientists in their respective disciplines. Therefore, while the Chair position provided a superb platform to make a real difference for women in science and engineering, it also came with a personal and professional cost. For some of the Chairholders of the first generation, it was now time to pass the torch.[64]

Still, the first five Chairs provided a substantial platform for the next generation to build. They developed a database on participation rates and identified reliable

[63] Carolyn Emerson, Interview with Hannah Elizabeth Young, telephone interview, January 3, 2018; Elizabeth Cannon, Interview with Hannah Elizabeth Young, telephone interview, May 3, 2018; Claire Deschênes, Interview with Hannah Elizabeth Young, telephone interview, March 13, 2018; and Claire Deschênes, *Bilan final du second mandat de la Chaire CRSNG/Alcan pour les femmes en sciences et génie au Québec*, Québec, Université Laval, janvier 2007, p. 12.

[64] Barbara Muir, NSERC, "The NSERC/Industry Chairs for Women in Science and Engineering: A National Program in Canadian Universities", paper presented to 12th International Conference of Women Engineers and Scientists, Ottawa, July 2002, accessed November 28, 2018, https://www.mun.ca/cwse/icwes_cwse5.pdf; Carolyn Emerson, Interview with Hannah Elizabeth Young, telephone interview, January 3, 2018; Elizabeth Cannon, Interview with Hannah Elizabeth Young, telephone interview, May 3, 2018; and Claire Deschênes, Interview with Hannah Elizabeth Young, telephone interview, March 13, 2018.

data sources. The Chairs compiled a common knowledge base of research on factors that influence participation and produced Canadian benchmarks for some key international research trends. Across the considerable variety of the Chair programs, there was useful practical experience and information on program impact. Finally, and possibly more importantly, there was a much greater awareness about the importance of increasing the participation of women in science and engineering in universities across Canada than in the previous decade.[65]

[65] Barbara Muir, NSERC, "The NSERC/Industry Chairs for Women in Science and Engineering: A National Program in Canadian Universities", paper presented to 12th International Conference of Women Engineers and Scientists, Ottawa, July 2002; accessed November 28, 2018, https://www.mun.ca/cwse/icwes_cwse5.pdf

Chapter 3
The Second Generation of Regional Chairs (2003–2011)

Starting in 2003, a new generation of five women gradually took up the role of regional Chairs: as Atlantic Chair, Dr. Cecilia Moloney, a professor in the Department of Electrical and Computer engineering at Memorial University of Newfoundland, who was sponsored by Petro-Canada; as Quebec Chair, Dr. Nadia Ghazzali, a professor in the Department of Mathematics and Statistics at Université Laval, funded by Industrielle Alliance; as Ontario Chair, Dr. Valerie Davidson, professor in the School of Engineering at the University of Guelph, sponsored first by Hewlett-Packard Canada and, for her second term, by Research in Motion (RIM); as Prairie Chair, Dr. Julita Vassileva, a professor in the Computer Science Department at the University of Saskatchewan, funded by Cameco; and lastly, as the West Coast Chair, Dr. Anne Condon, a professor in the Department of Computer Science at the University of British Columbia, sponsored by General Motors Canada.

This second generation benefited from the fact that the first five Chairholders had established a strong reputation for the program and were outstanding role models. In fact, several of the first five regional Chairs served as mentors to the next generation. For example, Maria Klawe was engaged closely with Anne Condon as a member of her advisory committee while Mary Williams, a mentor and a personal friend to Valerie Davidson, encouraged her to trust her own instincts in developing a unique CWSE program for Ontario. Claire Deschênes worked hand in hand with Nadia Ghazzali during the transition between the two Quebec Chairs. She shared her files, and Nadia was able to hire Claire's assistant.[1]

[1] Cecilia Moloney, Interview with Hannah Elizabeth Young, telephone interview, August 15, 2018; Anne Condon, Interview with Hannah Elizabeth Young, telephone interview, March 12, 2018; Nadia Ghazzali, Interview with Hannah Elizabeth Young, telephone interview, July 3, 2018; Valerie Davidson, Interview with Hannah Elizabeth Young, telephone interview, February 3, 2018; and Anne Condon, Interview with Ariane Zamin, telephone interview, 2017. See also Valerie Davidson, "Outside of our Rooms: Networking Across the Academy", *Royal Society Conference*, 2007, p. 5, Accessed April 16, 2019, http://hdl.handle.net/10214/2798

C. D'Amours et al., *The Chairs for Women in Science and Engineering Program*, Women in Engineering and Science, https://doi.org/10.1007/978-3-031-54085-1_3

There were still some challenges in the transition between the two generations of Chairholders. Indeed, at the time of Valerie Davidson's appointment as the Ontario Chair in 2003, Claire Deschênes was the only other Chairholder because there were vacancies in three of the five regions. Gaps in regional representation across Canada continued until 2005 due to the time required for competitions, difficulty in securing industrial support, and delays in announcing new awards. The lack of continuity proved to be detrimental for sustaining networks and communication efforts, and it was challenging to develop efforts at a national level.

Consolidation of the National Network

Recognizing that national collaboration required additional resources beyond the support for individual regional programs, Valerie Davidson initiated efforts as early as 2003 to get funding for the development of a formal research alliance at the national level.[2] She brought together a team that included the five regional Chairholders as well as social scientists and advocates for women in science and engineering. This group was successful in applications to two Social Sciences and Humanities Research Council of Canada (SSHRC) programs: a 2005 Research Development Initiative (RDI) award and a 2007 Letter of Intent to the Community University Research Alliance (CURA) program. Both research projects were supported by NSERC and partnerships with Canadian Coalition for Women in Science, Engineering, Technology and the Trades (CCWESTT), several deans of science and engineering, and various senior administrators.[3] Although the final application to the CURA program was not funded, the SSHRC research initiatives demonstrated the value of national-level research to address the issues of underrepresentation of women in science and engineering as well as a formal network for the CWSEs.

With the success of these initial joint research grants, the NSERC program officer for the CWSE program proposed that NSERC support the creation of an official network to help the regional Chairs continue to collaborate and communicate as a national group. This initiative was not unusual for NSERC since a similar program, the Chairs in Design Engineering, already had funding that allowed the Chairholders to network. NSERC's decision to provide funds for the National Network was a

[2]Valerie Davidson, "Outside of our Rooms: Networking Across the Academy", *Royal Society Conference*, 2007, p. 5, Accessed April 16, 2019, http://hdl.handle.net/10214/2798; Valerie Davidson, Interview with Hannah Elizabeth Young, telephone interview, February 3, 2018; Nadia Ghazzali, Interview with Hannah Elizabeth Young, telephone interview, July 3, 2018; and Valerie Davidson, Ann Holmes, and associates, *Highlights of the CWSE-ON Program (2003–2011)*, Guelph, CWSE-ON, 2012, p. 9.

[3]Valerie Davidson, Interview with Hannah Elizabeth Young, telephone interview, February 3, 2018; "Chairs for Women in Science and Engineering Program", December 1, 2017, accessed April 16, 2019, http://www.nserc-crsng.gc.ca/Professors-Professeurs/CFS-PCP/CWSE-CFSG_eng.asp and Valerie Davidson, "Outside of our Rooms: Networking Across the Academy", *Royal Society Conference*, 2007, p. 6, Accessed April 16, 2019, http://hdl.handle.net/10214/2798

turning point in improving collaboration and coordination of the regional Chairs program at the national level. Since its official inception in September 2006, the network has facilitated interactions among the five Chairs in order to increase the effectiveness of the regional programs through shared information and resources. The network support allowed the regional Chairs to have regular face-to-face meetings and to undertake research activities that support the common objectives of the CWSE Program.[4] Elizabeth Croft and the next generation of Chairholders used this network as a foundation to build future successful applications to SSHRC.

Between 2006 and 2011, the regional Chairs worked together on several national initiatives. Among these, the regional Chairs published regular newsletters describing regional and network activities and held biannual meetings—one by teleconference and one face-to-face at a relevant conference—where Chairholders presented updates on their regional programs and discussed ideas for collaborative projects or studies. In addition, they created, in collaboration with the University of Guelph library, a digital repository for network materials such as newsletters, statistics related to the participation of women in science and engineering in Canada, tools, and research findings.[5] The Chairs also organized special events at conferences. For example, they led sessions where they presented updates on their activities and thematic workshops at multiple CCWESTT conferences. Valerie Davidson, a member of the Canadian Committee of Women in Engineering, helped Monique Frize organize the CCWE+20 workshop held in Ottawa in 2011 and all the Chairs participated in the workshop. The regional Chairs were invited as a group to participate in the Royal Society of Canada National Conference "Rooms of Their Own" in Edmonton in 2007 and the NSERC Summit in November 2010. In addition, the Chairs presented joint papers on some of the network research findings at the Prometea Conference in France in 2009 and the ICWES 15 Conference in Australia in 2011. The network also facilitated liaisons with national organizations such as CCWESTT and Engineers Canada and international bodies like International Network of Women Engineers and Scientists (INWES). The National Network was also a platform to assert concerns about the lack of diversity in the first group of recipients of the prestigious Canada Excellence Research Chairs (CERC) in 2010. Not one woman was included in the group of 19 researchers who were selected to receive a CERC award which provided up to $10-million in total in federal money. Supported by the statistical analysis that Nadia Ghazzali had conducted for the Canada Research Chair (CRC) and CERC programs, the network of Chairs wrote a letter to the Industry Minister, Tony Clement, to express their concerns about structural biases facing female researchers. They asked for a review of the search and

[4] Valerie Davidson, Interview with Hannah Elizabeth Young, telephone interview, February 3, 2018; "Chairs for Women in Science and Engineering Program", December 1, 2017, accessed April 16, 2019, http://www.nserc-crsng.gc.ca/Professors-Professeurs/CFS-PCP/CWSE-CFSG_eng.asp and Valerie Davidson, "Outside of our Rooms: Networking Across the Academy", *Royal Society Conference*, 2007, p. 6, Accessed April 16, 2019, http://hdl.handle.net/10214/2798

[5] Digital repository for National Network of Regional Chairs for Women in Science and Engineering can be accessed https://atrium.lib.uoguelph.ca/xmlui/handle/10214/2695

nomination processes and a clear commitment to ensuring equity and diversity in the CERC program.[6]

Valerie Davidson, the Ontario Chair, initiated the idea of a national alliance, but all of the regional Chairs agreed with the vision and were active contributors to the joint research projects. Between 2005 and 2011, the regional CWSEs undertook a number of joint research projects which are archived in the Guelph digital repository (Atrium) (https://atrium.lib.uoguelph.ca/xmlui/handle/10214/2732). For example, over 1,200 final-year undergraduate students in engineering and computer science programs from across Canada completed a 2011 survey, which assessed factors that draw students to engineering and computer science programs; factors that contribute to positive academic experiences; and students' plans following graduation from undergraduate programs and students' perceptions of workplaces. While some of the survey results were not differentiated by gender, the regional Chairs found some notable distinctions between men and women students which can inform recruitment and retention efforts. In terms of recruitment to undergraduate programs, interest in the various fields was important for women students in engineering but not as important as it was for their male counterparts. Women engineering students also cited more frequently encouragement from family and friends and the sense of being able to make a difference as reasons for choosing their field of study. In terms of educational experiences, overall, respondents described their experiences at similarly positive rates. However, for women students, support from instructors, staff, and teaching assistants was very important. A significantly higher proportion of women students agreed with the importance of having support from family and friends. In terms of the workplace climate, a majority of women students are expected to face gender-based discrimination in the workplace. The findings were presented at the ICWES 15 conference in Adelaide, Australia, and also shared with the National Council of Deans of Engineering (NCDEAS).[7]

Another joint research project focused on equity indicators for women faculty in science and engineering based on Statistics Canada data. This study was based on statistical analysis of the Survey of Earned Doctorates between 2003 and 2007 and

---

[6] Julita Vassileva, *Final Report, 2005–2011*, Regina, University of Saskatchewan, May 2012, p. 14; "National Network of Regional Chairs for Women in Science and Engineering", Date unknown, accessed April 16, 2019, https://atrium.lib.uoguelph.ca/xmlui/handle/10214/2695; "Report to CCWE+20", June 19, 2011, Accessed April 16, 2019, https://atrium.lib.uoguelph.ca/xmlui/handle/10214/2793; Valerie Davidson, Ann Holmes, and associates, *Highlights of the CWSE-ON Program (2003–2011)*, Guelph, CWSE-ON, 2012, p. 10 and Valerie Davidson, Interview with Ariane Zamin telephone interview, August 21, 2017.

[7] Valerie Davidson, Elizabeth Croft, Julita Vassileva, and Nadia Ghazzali, "Summary Report—2011 Career Transitions Survey", April 30, 2012, accessed April 16, 2019, https://atrium.lib.uoguelph.ca/xmlui/handle/10214/3537?show=full and Valerie Davidson, Ann Holmes, and associates, *Highlights of the CWSE-ON Program (2003–2011)*, Guelph, CWSE-ON, 2012, p. 1 et 10. Jennifer Fender, Valerie Davidson, Julita Vassileva, Nadia Ghazzali, and Elizabeth Croft, "Perceptions and Experiences of the Workplace Among Canadian Computer Science and Engineering Students—A Gender Analysis", paper presented to the *ICWES 15*, Adelaide (Australia), July 19–22, 2011, accessed April 16, 2019, https://atrium.lib.uoguelph.ca/xmlui/handle/10214/3520?show=full

the University and College Academic Staff System data collected between 2004 and 2007. This example shows the synergy that was developed across regional and national activities. Anne Condon's involvement with the Assessment Advisory Committee of the Faculty of Science at the University of British Columbia informed the development of relevant equity indicators, and the network project provided a broader, national-level context for the University of British Columbia (UBC) Faculty of Science review. The analysis showed that half of men faculty members in major disciplines in science and engineering were full professors, while the proportion of women was closer to one quarter. In terms of salaries, women faculty members for the academic year 2005–2006 earned between 87% and 90% of men's salaries. One reason explaining this situation was the higher concentration of women at lower academic ranks than men. However, the Chairholders found that greater earning gaps were still evident at the full professor level than at lower ranks. In addition, the study revealed that women continued to be underrepresented both as graduate students and in new faculty appointments in these disciplines. These results indicated that science and engineering faculties still had problems attracting and retaining female graduate students. The authors also suggested that these problems could not be tackled by relying on affirmative hiring practices alone. Furthermore, women still formed a small proportion of the doctoral graduates in engineering, computer science, and mathematics (ranging from 17% to 34%). With respect to career progress, although few differences were seen in time from hire to promotion between women and men, differences were still noted in tenure status with women being more likely to be untenured compared to men. Finally, regarding attrition from academic positions, the Chairholders showed higher attrition rates for women at the assistant professor rank for agricultural and biological sciences, mathematics, and physical sciences.

NSERC's decision to fund the National Network facilitated the Chairholders' ability to do collaborative work. As a group, they took up new projects, projects that might not have been possible before. The Chairs had a significant impact not only in their own respective regions but also nationally now. The consolidation of the network made a huge difference in the program at a national level. On a more personal level, the closer relationships between the Chairs provided an opportunity to share personal experiences and feelings or firsthand information. As for the first generation of Chairholders, some of the second generation of Chairs felt isolated. Thus, having a group of persons sharing the same position and facing the same challenges was more than welcomed by these five women.[8]

[8]Valerie Davidson, Ann Holmes, and associates, *Highlights of the CWSE-ON Program (2003–2011)*, Guelph, CWSE-ON, 2012, p. 1; Trudy Smit Quosai, Valerie Davidson, Nadia Ghazzali, Cecilia Moloney, and Julita Vassileva, "Defining Equity Indicators for Benchmarking Women's Participation in Science and Engineering Faculties across Canada", paper presented to the *Prometea Conference*, France, November 26–28, 2009, accessed April 17, 2019, https://atrium.lib.uoguelph.ca/xmlui/handle/10214/2780?show=full and Cecilia Moloney, Interview with Hannah Elizabeth Young, telephone interview, August 15, 2018.

The Atlantic Chair: Cecilia Moloney (2004–2009)

When Cecilia Moloney was appointed as the NSERC/Petro-Canada Chair for Women in Science and Engineering for the Atlantic Region (CWSEA) in July 2004, she had been a professor of electrical and computer engineering at Memorial University of Newfoundland for 14 years. Her research focused on nonlinear signal and image processing methods, human visual perception, radar signal processing, and signal representations. Prior to her years as a faculty member at Memorial, she had worked for three years as a member of the scientific staff at Bell-Northern Research in Ottawa and for two years with INRS-Télécommunications of the Université du Québec in Verdun.[9] As the only female faculty member in her department for almost a decade, she understood the hardships faced by women in science and engineering fields. Yet, as for numerous other Chairholders, the creation of the National Chair and the Montreal Massacre were two events that increased her own awareness about women in science and engineering issues. Her involvement in NSERC's committee for the University Faculty Awards and Memorial's committee for the first round of applications for the NSERC Chair in 1996 also strengthened her interest in these matters.[10]

Cecilia's work as the Atlantic Chair was influenced by strong support in three areas. As was the case for Mary Williams, the presence of several strong organizations like the Newfoundland and Labrador Women in Science and Engineering (WISE) chapter, the Newfoundland-based Women in Resource Development Corporation (WRDC), and the association of Professional Engineers and Geoscientists Newfoundland and Labrador (PEG NL) gave her the opportunity to build strong partnerships. Second, Petro-Canada, the industry sponsor of the Chair,

[9] "Cecilia Moloney", July 8, 2015, accessed March 27, 2019, https://www.mun.ca/engineering/about/people/ceciliamoloney.php and "Cecilia Moloney", September 30, 2010, accessed March 27, 2019, https://www.mun.ca/msru/members/cmoloney.php

[10] Cecilia Moloney, Interview with Hannah Elizabeth Young, telephone interview, August 15, 2018.

was actively engaged throughout Cecilia's term. In this regard, Petro-Canada provided useful feedback in response to the first progress report submitted after 24 months of activity. Although Petro-Canada found that on the whole, the objectives of the Chair were being met, their representatives on the Advisory Board provided feedback on the report that changed the direction of the Chair and helped in the development of the program for the last three years. The third major influence on the Chair's work was the creation of the National Network in 2006. Initially, research was not a major thrust of Cecilia's program. The cross-Canada group of researchers from both NSERC and SSHRC disciplinary backgrounds established through Ontario Chair Valerie Davidson's lead gave Cecilia a better understanding of the importance of research as a prelude to the design and implementation of effective programming. She increasingly included more such research in her own program after 2008.[11]

The activities of the Atlantic Chair were divided into six general areas: outreach strategies, communication, networking, research, role model, and teamwork. Among the programs developed by Cecilia, three really stand out. The first key activity was the direct outreach to graduate students. Following up on the workshop *The Graduate Student Leader*, developed for presentation at the 2006 CCWESTT Conference, Cecilia encouraged the formation of a Graduate Student Discussion Group at Memorial University in the fall of 2006. Initially, the group consisted of graduate students who had attended a dry run of the CCWESTT workshop at Memorial prior to the conference. The group met every other week during the fall and winter semesters to discuss leadership and lab management and by March 2007 started to plan their own workshop on leadership for the 2008 CCWESTT Conference. Through partial financial support from the Chair, five members of this group developed a 50-min workshop "Women as Leaders, Mentors, Scientists/ Engineers," successfully presented in a multimedia format with facilitated breakout sessions. In fall 2009, the group connected itself to WISE Newfoundland and Labrador, thus reviving the WISE Graduate Student Section (GSS) which had become inactive after 2005. As the WISE GSS, and in conjunction with the Chair, the group developed a workshop for a membership drive event at Memorial University, entitled "To 'Rapport' or not to 'Rapport': How to make people like you in 90 s."[12]

Cecilia focused on networking with a special emphasis on Atlantic Canada in order to maintain and strengthen the Atlantic communication network through which the regional nature of the CWSE would be realized. She undertook a tour of Atlantic Canada in fall 2006, as well as other Atlantic Canada travel between 2006 and 2009. The CWSEA Advisory Board meeting of 2007 brought together representatives from the four Atlantic provinces to discuss collaboration across the region. Through the National Network, Cecilia became very active across the

[11] Cecilia Moloney, *Final progress report submitted to Natural Sciences and Engineering Research Council of Canada*, St. John's, Memorial University, May 2009, p. 15–16.

[12] Moloney, *Final progress report*, 5.

country. In 2008, she attended the colloquium "Les grands enjeux des femmes pour un développement durable: démocratie, citoyenneté et leadership," held at l'Université de Moncton, co-hosted by the Quebec-based AFFESTIM. She was also an active participant in the CCWESTT conference held in Guelph in 2008, providing input into the planning of two sessions, two presentations, and partial support for travel for nine students. CWSEA also networked at other national and international events, including: as session Chair at an NSERC-organized Organisation for Economic Co-operation and Development (OECD) Workshop on Women in Science, Engineering and Technology, held in Ottawa on September 2006; as presenter in a CWSE session at a Royal Society of Canada organized conference in Edmonton in May 2007; as presenter at ICWES 14 in France in July 2008; and as an invited presenter to the Science, Technology and Management Group of the Canadian Space Agency, in December 2008.[13]

CWSEA developed strategies and activities which were informed by the geographic, economic, and sociocultural conditions of Atlantic Canada and implemented with the aid of collaborative links across the region. Cecilia's other research interests (and her activities as the CWSEA) were located within gender and science/engineering studies, with a specific focus on human knowing and agency, innovative pedagogy, social innovation, and strategies for change. In October 2005, Cecilia gave a CWSEA public lecture at Memorial entitled, "Women doing science in a Nelson man-of-war," in which she compared the structure of the current scientific enterprise with the organizational structure aboard a naval man-of-war of the Nelson era.[14]

Cecilia considers her biggest legacy to be the *MetaKettle Project*. This project originated from the work done for the SSHRC application to support the National Network. In July 2004, she attended the first formal brainstorming meeting which led to the network and was an active member in the network throughout her term. The network was successful in obtaining a SSHRC Research Development Initiative (RDI) grant (2005–2007). A portion of the RDI grant was used to develop an application for an SSHRC Community University Research Alliance (CURA) in the 2007 competition. Although the CURA application was unsuccessful, the connections formed by the network and the insights and knowledge obtained in developing the application had not been wasted. In meetings of the National Network of CWSEs, the Chairs discussed how they might continue their joint work, by securing funds in smaller amounts for selected topics of research, for example. At the same time, the National Network of CWSEs continued with smaller projects funded by their common network funding.[15]

As a co-applicant for the CURA competition, each Chair had to develop a particular theme. Cecilia began to work on the impact of pedagogy in attracting and

[13] Moloney, *Final progress report*, 9–10.

[14] "Cecilia Moloney", September 30, 2010, accessed March 27, 2019, https://www.mun.ca/msru/members/cmoloney.php

[15] Moloney, *Final progress report*, 11.

retaining women into undergraduate studies in sciences and engineering. Over time, she was convinced that this specific aspect was crucial to the issue. Thus, toward the end of her tenure as the Atlantic Chair, she used Chair funds to hire an associate researcher to help her develop the idea of the *MetaKettle Project*.[16] Building upon the mandate of the Chair to increase the participation of women in science and engineering, the project broadened its goals. It aimed to "increase the diversity of students studying science and engineering, while enhancing the long-term sustainability of graduates over their careers and lives."[17] The *MetaKettle Project* was in fact a response to two key problems evident within engineering and science education as well as within industry and organizations employing engineering and science graduates. It aimed to increase the diversity of students studying science and engineering while enhancing the long-term sustainability of graduates over their careers and lives. The *MetaKettle* umbrella included several projects. For example, Cecilia and her staff assisted the Faculty of Engineering and Applied Science in preparing for outcomes-based evaluation measures required for accreditation of their engineering programs in 2013. The team also developed module materials on leadership for graduate students, on how to think about science, and on engineering ethics for courses in the engineering profession stream. Cecilia's team collaborated with colleagues in the Faculties of Arts and Education on a research project on creative methods to foster research writing for graduate students in the humanities and in engineering. Finally, they engaged in research projects to obtain evidence-based results to inform the development of the *MetaKettle* program.[18]

The Quebec Chair: Nadia Ghazzali (2006–2014)

Prior to her appointment as the NSERC/Industrielle Alliance Chair for Women in Science and Engineering in Quebec in 2006, Nadia Ghazzali was already an accomplished mathematician and statistician as well as an experienced academic administrator. She joined the Department of Mathematics and Statistics at Université Laval in 1993. From 2001 to 2006, she successively held the position of Director of the undergraduate program in statistics, Vice Dean for Research and Development at the Faculty of Science and Engineering, and Associate Vice President for Research in the same institution. She was appointed as a member of the Academy of Science and Technology of Morocco in 2006. In addition to her various administrative functions, Nadia had made significant research contributions in several areas of mathematics and statistics, including discriminant analysis, classification methods, and artificial neural networks for modeling, and characterizing large databases derived

[16] Cecilia Moloney, Interview with Hannah Elizabeth Young, telephone interview, August 15, 2018.

[17] Janna Rosales and Cecilia Moloney, "The MetaKettle Project: A CWSEA Legacy Pedagogy Project: Brewing the Holistic Advantage", October 29, 2010, accessed March 27, 2019, https://www.engr.mun.ca/~cmoloney/MK/MetaKettleOverview.pdf

[18] Memorial University, "The MetaKettle Project", n.d., accessed March 27, 2019, https://mun.yaffle.ca/projects/2425

from the fields of signal processing, medical imaging, pattern recognition, astrophysics, and biostatistics.[19] Although she had been constantly pursuing questions about mathematics and statistics, Nadia was not oblivious to the discrimination faced by women in her professional field. As a young girl, Nadia and her sisters had been pushed hard by their mother. Being illiterate and single, her mother struggled to raise her family. Thus, she wanted her daughters to study and become self-sufficient women. Her mother's "feminist" perspective of life proved instrumental not only to Nadia's desire to engage in her mathematics studies but also to her drive to improve women's status in science and engineering and to becoming a role model for young girls later in life.[20]

As in the case of Claire Deschênes, the former Quebec Chair, the province's distinct setting had a strong influence on Nadia's work. At the time of her appointment in 2006, and throughout her tenure, the Quebec government's continuing concern with the skilled labour force proved to be a beneficial stimulus for her efforts to attract women to science and engineering. Thus, Nadia found a fertile ground for her initiatives at all levels of the government apparatus. For example, for her research projects on Quebec enrollments at undergraduate and graduate levels and the status of women professors in science and engineering faculties, she was given access at once to the Ministry of Education's data. Another factor that positively influenced her tenure as Chair was the interest of the francophone mainstream media. For example, for the 100-year anniversary of Marie Curie's Nobel Prize in Chemistry in 2011, Radio-Canada asked Nadia to reflect on Curie as a role model. It seemed to her to be much harder for her fellow NSERC Chairholders to get the attention of the media in other parts of Canada. Finally, the presence of the Marianne-Mareschal Chair for the promotion of engineering among women at Polytechnique Montréal since 1998 gave another impetus to the Laval Chair. Indeed, Nadia worked with the

[19] "Nomination de Nadia Ghazzali à titre de rectrice de l'Université du Québec à Trois-Rivières", January 20, 2012, accessed March 28, 2019, https://www.uquebec.ca/reseau/fr/medias/communiques-de-presse/nomination-de-nadia-ghazzali-titre-de-rectrice-de-luniversite-du-quebec-trois-rivieres and "Nadia Ghazzali", November 21, 2012, accessed March 30, 2019, https://www.scienceadvice.ca/experts/nadia-ghazzali/

[20] Nadia Ghazzali, Interview with Hannah Elizabeth Young, telephone interview, July 3, 2018. See also "Nadia Ghazzali: Professeure de statistiques à l'Université Laval", *Le Soleil*, June 16, 2008, p. 28.

Marianne-Mareschal Chairholders to take existing initiatives and set them up in the Quebec City urban area. This crucial collaboration lasted for her whole tenure as Chair, thus providing a formidable ally for Nadia to change Quebec's scientific culture.[21]

Nadia undertook three vital types of activities during her time as Quebec Chair: research, promotion, and international collaboration. Among the analyses she has done to understand the underrepresentation of women in science and engineering in Quebec, three statistical studies clearly stand out. As a statistician herself, it was important to prove, not with a sample but with comprehensive data, that there was actually a real shortage of women in science and engineering in the province. Using data from the Ministry of Education, Nadia was able to show that women remained a minority in the natural sciences (pure and applied) and engineering. In 2007, they represented only 12% of the undergraduate students in these fields. At the undergraduate level, the health sciences sector attracted more and more women students, at the expense of the science and engineering sectors, she argued. In these fields, there were only 8,396 women in Quebec in 2007 compared to 9,943 in 1999, a decrease of 16% over eight years.[22]

A second project involved the statistical analysis of science and engineering faculty members in Quebec. Using data from all 18 universities of the province, the goal was to draw a detailed portrait at the faculty level between 1997 and 2007. The report highlighted the fact that, despite an increase in the number of women between 1997 and 2007, the gender composition remained at 16% in 2007 for all areas of science and engineering. Similarly, regarding the evolution of promotions, while the majority of full professors were men, more women professors had the status of assistant or associate professor. As a result, the average income of women professors was lower than that of their male counterparts, mainly because there were more men full professors, which helps to explain this difference in the average salaries.[23] Linked to the status of Quebec faculty, Nadia undertook a third statistical analysis project to document the glass ceiling in scientific research. In the wake of the federal government's awarding of the first 19 CERCs in the spring of 2010, all of which

[21] Nadia Ghazzali, Interview with Hannah Elizabeth Young, telephone interview, July 3, 2018. See also "'Un dimance avec Marie Curie' à Radio-Canada le 4 décembre", December 3, 2011, accessed March 30, 2019, http://www.lienmultimedia.com/spip.php?article30063

[22] Véronique Boiteau and Nadia Ghazzali, "Profils des inscriptions des étudiants aux trois cycles universitaires au Québec de 1999 à 2007", NSERC/Industrielle Alliance Chair for Women in Science and Engineering in Québec, 2008 and Nadia Ghazzali, "La situation des femmes dans les milieux académique et professionnel au Québec", *Le SPUL-lien: Le bulletin socioprofessionnel du Syndicat des professeurs et professeurs de l'Université Laval*, vol. 7, no. 1 (March 2011), p. 5.

[23] David Émond et Nadia Ghazzali, *Analyse du corps professoral universitaire en sciences et génie au Québec, de 1997 à 2007*. Chaire CRSNG/Industrielle Alliance pour les femmes en sciences et génie au Québec, 2009; Nadia Ghazzali, "La situation des femmes dans les milieux académique et professionnel au Québec », *Le SPUL-lien: Le bulletin socioprofessionnel du Syndicat des professeurs et professeurs de l'Université Laval*, vol. 7, no. 1 (March 2011), p. 5 and Nadia Ghazzeli, *Rapport final du second mandat de la Chaire CRSNG/Industrielle Alliance pour les femmes en science et génie*, Quebec, Université Laval, 28 mai 2012, p. 6.

were awarded to men, she decided to do a comparative analysis of 1,800 Canada Research Chairs and the new Canada Excellence Research Chairs using data from the Canada Research Chair Secretariat. According to her study, the proportion of all types of Chairs (from all academic disciplines) awarded to women was roughly one quarter (25.2%). For senior Chairs, the proportion decreased further. In fact, out of 769 Chairs held by senior investigators, only 127 (16.5%) were held by women.[24]

As part of her outreach efforts, Nadia partnered with the Marianne-Mareschal Chair to set up two events in Quebec City. The first event, *Les filles et les sciences: un duo électrisant!*, was presented for the first time at Polytechnique Montréal in 1999. This activity aimed to inform participants about the more detailed nature of scientific and technological careers and to stimulate vocations. Through experimental workshops and discussions, young girls also had the opportunity to meet female role models working in science and technology fields. This event grew bigger and bigger in the following years, bringing together a total of 730 girls of Secondary 2 and 3 (the equivalent of Grades 8 and 9 elsewhere in Canada) in three separate locations (Montreal, Sherbrooke, and Rimouski) in 2007. Nadia understood that this event had tremendous potential to battle myths and bias associated with nontraditional careers in science and engineering. Together with three other female faculty members, she decided to organize the event at Université Laval. The first edition, presented in February 2008, attracted a total of 60 girls and their parents. Overall, the results from the survey distributed to participants, volunteers, companions, and adult participants were very positive. The increase in attendance during Nadia's time as Chair was also a testimony of its success, reaching an apex of 110 in 2010. As of 2019, *Les filles et les sciences: un duo électrisant!* is still a great success story, involving more than 150 participants for the event organized by Université Laval.[25]

The partnerships with the Marianne-Mareschal Chair did not stop there. The Quebec Chair provided funding for the program *Future ingénieure?* which targeted first-year Collège d'enseignement general et professionnel (CEGEP) students (the postsecondary level after high school providing general and vocational education before university). This specific period is critical in Quebec since most college

[24] Nadia Ghazzali and Nydia Morin-Rivest, *Chaires de recherche du Canada et chaires d'excellence en recherche du Canada: statistiques détaillées*. Chaire CRSNG/Industrielle Alliance pour les femmes en sciences et génie au Québec, 2010; Nadia Ghazzali, "La situation des femmes dans les milieux académique et professionnel au Québec », *Le SPUL-lien: Le bulletin socioprofessionnel du Syndicat des professeurs et professeurs de l'Université Laval*, vol. 7, no. 1 (March 2011), p. 5 and Nadia Ghazzeli, *Rapport final du second mandat de la Chaire CRSNG/Industrielle Alliance pour les femmes en science et génie*, Quebec, Université Laval, 28 mai 2012, p. 6.

[25] Suzanne Lacroix, *Rapport d'activités 2009–2010*, Montréal, Marianne-Mareschal Chair, 2010, p. 7–8; Nadia Ghazzali, Interview with Hannah Elizabeth Young, telephone interview, July 3, 2018; "Les filles et les sciences 2019: un duo électrisant", February 27, 2019, accessed March 31, 2019, https://www.uquebec.ca/reseau/fr/medias/actualites-du-reseau/les-filles-et-les-sciences-2019-un-duo-electrisant; Nadia Ghazzeli, *Rapport final du second mandat de la Chaire CRSNG/Industrielle Alliance pour les femmes en science et génie*, Quebec, Université Laval, 28 mai 2012, p. 4 and "Les filles et les sciences: un duo électrisant", *Le fil de l'événement*, vol. 43, no. 21 (February 14, 2008), accessed March 30, 2019, https://www.lefil.ulaval.ca/campus-express-5722/

students decide at the end of their first year which field they will study at university. Thus, according to their field of interest, the program matched young women to engineers to enable them to better understand the engineering profession. Two types of visits were offered to the students: individual and exploratory. As part of an individual visit, the student became the right hand of an engineer for one day. As part of an exploratory visit, groups of students met with a group of engineers from different specialties. The program was first started in the Quebec City area in 2008. For the 2010 edition, the 12 participants were sponsored by six companies and research centers: Université Laval's Water Engineering Research Group, the Defence Research and Development Canada in Valcartier, Sainte-Justine Hospital in Montreal, the Quebec Ministry of Transport, the engineering consulting firm Dessau, and the multinational company Canam group.[26]

In addition to her outreach activities, Nadia worked closely with a number of national and international organizations. In 2011–2012, she was asked to be part of the Expert Panel on Women in University Research of the Council of Canadian Academies. The panel's mandate, which focused on the issue of women in academia, was to provide independent scientific assessments to inform public policy and decision-making. Thus, she was part of the team behind the publication in 2012 of the report *Strengthening Canada's Research Capacity: The Gender Dimension* which addressed policies and societal, cultural, institutional, and economic factors influencing the career trajectories of women researchers in Canadian universities. The report also underlined the gender disparities observed in the statistical profile of Canadian university researchers.[27] At the international level, Nadia quickly became involved with INWES. Since the organization's headquarters were situated in Ottawa, she felt she needed to lend a hand. As the secretary general from 2008 to 2011, she organized four board of directors' meetings which took place during the INWES Regional Conferences in South Korea, the United States, and Australia. It was also in this context that she was involved in the preparation of the seventh General Assembly of INWES members. Nadia also contributed to the establishment of INWES Arab World, one of INWES's regional networks, and the Moroccan Association of Engineers and Scientists.[28]

[26] Suzanne Lacroix, *Rapport d'activités 2009–2010*, Montréal, Marianne-Mareschal Chair, 2010, p. 7–8; Nadia Ghazzali, Interview with Hannah Elizabeth Young, telephone interview, July 3, 2018, and Nadia Ghazzeli, *Rapport final du second mandat de la Chaire CRSNG/Industrielle Alliance pour les femmes en science et génie*, Quebec, Université Laval, 28 mai 2012, p. 4.

[27] "L'Oréal Canada for Women in Science", unknown date, accessed March 31, 2019, https://www.loreal.ca/csr-commitments/the-l'oréal-corporate-foundation/science/l%27oréal-canada-for-women-in-science; Nadia Ghazzali, Interview with Hannah Elizabeth Young, telephone interview, July 3, 2018, and Nadia Ghazzeli, *Rapport final du second mandat de la Chaire CRSNG/Industrielle Alliance pour les femmes en science et génie*, Québec, Université Laval, 28 mai 2012, p. 15–16.

[28] "Nomination de Nadia Ghazzali à titre de rectrice de l'Université du Québec à Trois-Rivières", January 20, 2012, accessed March 28, 2019, https://www.uquebec.ca/reseau/fr/medias/communiques-de-presse/nomination-de-nadia-ghazzali-titre-de-rectrice-de-luniversite-du-quebec-trois-rivieres and Nadia Ghazzeli, *Rapport final du second mandat de la Chaire CRSNG/*

Before the end of her Chair term, Nadia was appointed President ("Rectrice") of Université du Québec à Trois-Rivières, where she continued her Chair work until 2014.

The Ontario Chair: Valerie Davidson (2003–2011)

At the time of her appointment as the NSERC/Hewlett-Packard Canada Chair for Women in Science and Engineering for the Ontario Region (CWSE-ON) in 2003, Valerie Davidson had been a professor at the School of Engineering of the University of Guelph for 15 years. Being interested in the interface between engineering and food, the southwest Ontario institution provided a stimulating research environment for her since it covered the "whole spectrum from food production to processing to the hospitality side."[29] There she was able to work on projects related to food safety, more specifically on risk assessment for microbial hazards in foods and risk prioritization based on public health impact, consumer perception and acceptance of risk, potential market impact, and social factors.[30] As for several of her fellow Chairholders, the Montreal Massacre of 1989 touched her deeply. She recognized many similarities between the victims and the female students she was teaching at the University of Guelph, and this reinforced her sense of the terrible loss of talented engineers and community leaders. She was already involved as a member of the national executive of WISE and agreed to serve as a member of the Canadian Committee on Women in Engineering (CCWE), which traveled the country and investigated the state of women in the profession from 1990 to 1992 and ultimately

---

Industrielle Alliance pour les femmes en science et génie, Quebec, Université Laval, 28 mai 2012, p. 16–17.

[29] "Big Ideas: Blazing a trail for female engineers", May 25, 2018, accessed March 31, 2019, https://www.eng.mcmaster.ca/news/big-ideas-blazing-trail-female-engineers

[30] "Research", unknown date, accessed March 31, 2019, http://www.valdavidson.ca

published the *More Than Just Numbers Report*. After the CCWE report was finished, she focused on obtaining tenure and promotion to full professor while she continued to offer support for women students in science and engineering. For example, she established a successful mentoring program to encourage Guelph students to apply for National Research Council of Canada (NRC)'s Women in Engineering and Science program. When the 2002 call for applications for the CWSE program was announced, the Dean of the College of Physical and Engineering Sciences proved to be an enthusiastic ally, agreeing to secure financial support from an industrial sponsor. Mary Williams, who had held the Atlantic Chair from 1997 to 2002 and was a good friend of Valerie's as well as a role model, encouraged her to apply for the Ontario Chair position. This final push proved to be instrumental in her decision to apply.[31]

Valerie's work as the Ontario Chair was shaped by a number of factors. First, she was a professional engineer who also had a graduate degree in food science, and many of her research projects were based on interdisciplinary collaborations. The integration of science and engineering in the Chair program was facilitated by the fact that the Chair was supported and reviewed at the level of the College of Physical and Engineering Sciences (Physics, Chemistry, Mathematics, and Engineering). Finally, Valerie was able to obtain additional funding for the Chair program, but most funds were tied to youth outreach activities. For example, in 2003, the Ontario Women's Directorate asked Valerie to lead the Science and Technology Consortium which included WISE Sudbury, CCWESTT, Lakehead and Ryerson universities, Actua, and the Chair for Women in Science and Engineering—Ontario (CWSE-ON). The consortium members shared resources for outreach activities and Ryerson and the CWSE-ON co-developed *Ready SET*[32] *Go!* workshops for classroom delivery. Research in Motion (RIM), the sponsor of Valerie's renewal term (2008–2011), was actively engaged in youth outreach. Consequently, the agreement to fund the second term included target numbers for outreach programs.[33]

Ontario, with 15 schools or faculties of applied science and engineering, was the province with almost half of the undergraduate engineering students in all of Canada. At the time that Valerie became the Ontario Chairholder, there was concern that, despite efforts to encourage more women to study engineering, there were signs that participation was declining. In February 2005, Tom Harris, Dean of the Faculty of Engineering and Applied Science at Queen's University, and Valerie co-hosted a roundtable discussion with representatives from all faculties/schools of applied science and engineering in Ontario. There was unanimous agreement that it was critical to work collectively to provide opportunities for young women to

[31] Valerie Davidson, Interview with Hannah Elizabeth Young, telephone interview, February 3, 2018; "A day of reflection", December 6, 2016, accessed March 31, 2019, https://blog.ospe.on.ca/community/2686/

[32] The SET acronym stands for Science, Engineering and Technology

[33] Valerie Davidson, Interview with Hannah Elizabeth Young, telephone interview, February 3, 2018; Valerie Davidson, Ann Holmes, and associates, *Highlights of the CWSE-ON Program (2003–2011)*, Guelph, CWSE-ON, 2012, p. 11.

explore engineering and to introduce them to a range of role models (engineering students, alumna, faculty, and staff). The Ontario Network of Women in Engineering (ONWiE) was formed as a platform to facilitate collaboration, networking, and sharing of resources and best practices. The roundtable group also decided to go forward with a province-wide event for girls in Grades 7–10 and the first *Go Eng Girl/GÉNiales, les filles* event was launched in October 2005. The event was advertised through central and local media releases and brought together 705 girls and 500 parents for a half-day of hands-on design projects led by engineering students and talks from keynote speakers about studies in different engineering disciplines. While the girls engaged in hands-on design activities in small teams, parents took part in an information session about the path to a career in engineering. The day concluded with a lunch, an information fair, and handouts such as T-shirts and an information sheet from all 15 engineering programs in Ontario.[34] *Go Eng Girl/GÉNiales, les filles* has been successful as an annual event since 2005 because of the strong commitment from its members, including a significant number of volunteers. Since 2005, the number of women students applying to engineering programs in Ontario has tripled.[35] As of 2018, the program involved 1,200 girls, 480 volunteers and organizers, and 25 ONWiE member schools from all across Canada on a yearly basis.[36]

As the Ontario CWSE, Valerie championed ONWiE's efforts and helped support ONWiE activities by providing financial and administrative resources (e.g., a website and communication materials). Each year, ONWiE activities are celebrated and recorded in an attractive brochure that captures images from events across the province (and beyond) and qualitative and quantitative indicators of impact. The report is shared with sponsors and supporters including the Council of Ontario Deans of Engineering (CODE). As Valerie's second term was coming to a close in 2011, she began to explore a new model for supporting ONWiE. She was able to build a convincing case for ONWiE's impact and the Ontario deans of engineering agreed to support an ONWiE Chair for a five-year period. Dr. Mary Wells (University of Waterloo) was the first Chairholder (2012–2017) and, Dr. Kim Jones (McMaster University) followed Mary as ONWiE Chair when the Ontario deans agreed to provide continuing support. At the time of this writing, Kim still holds the ONWiE Chair. Both Mary and Kim have been successful in expanding the network's impact and taking ONWiE in new directions. Mary, Kim, and Valerie recently published a

---

[34] Frize, *The Bold and the Brave*, p. 172 and ONWiE, *2005 Record of Achievement*, Guelph, ONWiE, 2005, accessed April 4, 2019, http://www.onwie.ca/wp-content/uploads/2018/07/2005-record-of-achievement.pdf

[35] Wells, M.A., Jones, K. and V.J. Davidson 2019. Ontario Network of Women in Engineering Case Study: Indicators of Success and Reflections on Lessons Learned, Special Issue of the International Journal of Gender, Science and Technology, 11 (1), 30–40.

[36] ONWiE, *2018 Report*, Guelph, ONWiE, 2018, p. 2, accessed April 4, 2019, http://www.onwie.ca/wp-content/uploads/2019/02/2018-Report-2-compressed.pdf; Valerie Davidson, Interview with Hannah Elizabeth Young, telephone interview, February 3, 2018, and Valerie Davidson, Ann Holmes, and associates, *Highlights of the CWSE-ON Program (2003–2011)*, Guelph, CWSE-ON, 2012, p. 4.

case study about ONWiE which highlights key successes as well as lessons learned about collaboration and effective outreach programs.[37]

Valerie and her team also delivered workshops and hands-on activities to encourage interest in the sciences and computer science (e.g., the *Codes, Computers, and Communicating with Aliens* workshops). The *Ready SET Go!* workshops were offered in 250 schools across Ontario, and 45% of the participants were students in rural and remote areas. Valerie also collaborated with the RIM team to develop a unique professional development workshop for elementary school teachers entitled *Science Made Simple*. The first workshop was held at the Perimeter Institute in 2010, and it included a number of simple hands-on activities that teachers could use to demonstrate concepts in the science curriculum. Workshop leaders focused on helping teachers gain confidence in assisting students to design and troubleshoot their prototypes.

Outreach activities were important to encourage young women to pursue post-secondary programs in science and engineering, but Valerie recognized that it was also important to address the retention of women as professionals in these fields. Valerie and her team developed a number of initiatives for female students in post-secondary programs, postdoctoral fellows, and faculty members that focused on networking, sharing leadership experiences and practical skills such as salary negotiations, and developing a skills portfolio. She also co-presented *Gender and Communication in the Engineering Faculty* workshops to faculty, staff, and administrators at two Ontario universities and posted the *Gender Issues Training Kit* based on the workshop materials on the website of the Ontario Society of Professional Engineers (OSPE).

A key component of Valerie's strategy was to conduct relevant research as a basis for understanding the environment for women in science and engineering and recommending changes. As noted in the discussion earlier in this chapter about the National Network, much of this research was designed and carried out with the other regional Chairholders. The *2010 Survey of Working Conditions* was sponsored by CWSE-ON in collaboration with OSPE, Professional Engineers Ontario, and Engineers Canada. It assessed issues related to mentoring, the impact of "unwritten rules" in the workplace, retention, experiences of perceived discrimination, and the availability, use, and impact of leaves. A decade later, it remains the most comprehensive snapshot of the engineering workplace in Canada.

Over her eight years as Ontario Chair, Valerie and the CWSE-ON team undertook a broad range of initiatives, and these have been summarized in "*Joining WiSE Conversations: Strategies and Successes from CWSE-ON 2003–2011*." This record as well as a number of resources that were created are archived in the atrium, which is a digital repository at the University of Guelph library (accessible at https://atrium.lib.uoguelph.ca/xmlui/handle/10214/2694). CWSE-ON initiatives are described for three distinct groups: elementary and secondary students, women in

[37](Wells, M.A., Jones, K. and V.J. Davidson 2019. Ontario Network of Women in Engineering Case Study: Indicators of Success and Reflections on Lessons Learned, Special Issue of the International Journal of Gender, Science and Technology, 11 (1), 30–40).

postsecondary programs in science and engineering, and women who work in science and technology professions. The majority of participants in CWSE-ON activities were women, but some activities were for men and women. Between 2003 and 2011, it is estimated that a total of 56,500 individuals (students, teachers, parents, and community leaders) participated in the CWSE-ON program.[38]

The Prairie Chair: Julita Vassileva (2005–2011)

When Julita Vassileva was appointed as the NSERC/Cameco Chair for Women in Science and Engineering for the Prairies in 2005, she had been a professor in the Computer Science Department at the University of Saskatchewan for six years. Her research involved human issues in decentralized software environments: user modeling and personalization, designing incentive mechanisms for encouraging participation, and facilitating trust in decentralized software applications, such as online communities, social networks, multi-agent systems, and peer-to-peer systems. Before moving to Canada in 1997 to be a postdoctoral fellow at the University of Saskatchewan, Julita did not experience the same bias against women in science and engineering that her fellow Canadian Chairs had. In Bulgaria, half of the mathematics program was filled with women. Yes, there was quota, but it was there to ensure place for men. Still, she must have felt bewildered when she first encountered gender discrimination at the University of Saskatchewan. Although it did not stop her in pursuing her career in Canada, the feeling must have stuck with her. Thus, when she was asked if she would be interested to take up Elizabeth Cannon's Chair, she accepted right away.[39]

---

[38] Valerie Davidson, Ann Holmes, and associates, *Highlights of the CWSE-ON Program (2003–2011)*, Guelph, CWSE-ON, 2012, p. 2.

[39] Julita Vassileva, Interview with Hannah Elizabeth Young, telephone interview, May 6, 2018, and "Dr. Julita Vassileva: Director", date unknown, accessed April 6, 2019, https://inwes-eri.weebly.

Julita set out to design her program based on her experiences and connections. She was inspired by the example of Elizabeth as Chairholder. Like the former Prairie Chair, Julita formed an Advisory Committee to include the main stakeholders and people whom she could use as contacts and points of leverage in important organizations. The success of the *CyberMentor* program served as a catalyst for Julita's *myWISEmentor* program which was launched at the University of Saskatchewan in 2009. Second, her own specialty and area of expertise directed her efforts. As a computer scientist researching online communities, social networks, and motivational strategies, several of her projects naturally aimed to connect people with interest in women in science and engineering in the Prairies, in addition to the creation of national and international networks. Cameco, the industry sponsor of the Chair, was actively engaged in the three final years of Julita's term. Indeed, Cameco had many women in high-level positions at the time and was generally considered a progressive company supporting its women. Thus, when Julita approached Cameco in 2008 with a proposal to start a program for Indigenous science outreach that would cost up to $5,000, the company accepted at once to sponsor the program entirely. Finally, as a result of the rapid growth of the Prairie provinces' Indigenous population, students of Indigenous descent were increasingly attending the regional universities. However, most Indigenous women students enrolled in programs like nursing, native studies, and law. Because Julita saw this as an opportunity to engage this community, she directed a sizable portion of her effort toward Indigenous peoples.[40]

The Prairie Chair focused her attention on three major areas: research, contributions to building communities and connecting women, and science outreach. In terms of research, she first looked at undergraduate students' attitudes toward science and the factors that influenced them to choose a science or engineering major. Using a sample of 250 undergraduate students enrolled at the University of Saskatchewan, Julita's team found a statistically significant difference in the factors influencing men and women students to choose a major and the reason behind the decision to pursue a graduate degree in these fields. For women students, the department climate was a predictor of their intentions of going into graduate school, while the availability of information about career options was a predictor for their male counterparts. A second project considered the provenance of Canadian women computer science faculty members and instructors. Since almost all Canadian women undergraduate students in computer science at the University of Saskatchewan did not pursue graduate studies, Julita wanted to know the provenance of the Canadian computer science department instructors. Having access to all faculty members and instructors registered in the Canadian Women in Computing Website database, she invited them to send an email stating where they completed their high school, undergraduate, and graduate studies. The results showed that about 45% came from

com/dr-julita-v.html

[40] Julita Vassileva, *Final Report, 2005–2011*, Regina, University of Saskatchewan, May 2012, p. 3–6 and 14–15.

Canada (where they had done their high school degree), an additional 15% came from the United States, and 40% came from outside of North America (mostly Western Europe). However, nearly 85% had done their doctoral degrees in North America.

A third project studied best practices to increase retention and improve workplace equity measures for women in Canada, the United States, and in Europe. Combining qualitative and quantitative data, Julita found that, in most Eastern European countries, women had reached parity in representation at undergraduate, graduate, and junior research/faculty levels in science and engineering for over three decades. However, she showed that high numbers of women in the pipeline do not result in automatic equality. Thus, she suggested that diversity policies need to focus on the conditions under which retention and advancement take place. A final project entailed acquiring and compiling gender-segregated statistics of enrollments of all the departments of science and engineering at the University of Saskatchewan between 2001 and 2007. Julita was able to show a consistent tendency of decline in the proportion of women's enrollment in fourth-year-level classes in these years, going from 15% in 2001–2002 to 8% in 2006–2007. She also found that women were more likely to enroll in chemical engineering while electrical and mechanical engineering were the two programs least likely to attract women, declining between 2002-2003 and 2006–2007 from 16% to 7% and from 16% to 3% respectively.[41]

With respect to her contributions to building communities and connecting women, the Prairie Chair undertook three main projects. Julita started the *Saskatchewan Science Network* (SSN) because she realized at the beginning of her mandate as Chair that there were already numerous science outreach organizations, educational initiatives by government organizations and grassroots movements that promote science and engineering to students. The Advisory Board stressed in January 2006 the need for better connections and networking among people working on similar initiatives. She decided to create a network modeled roughly after other science outreach networks such as the Alberta Science Network and the Calgary Science Network. Launched in 2008, the SSN served to coordinate efforts of all these organizations and thereby allowed them to share experiences, volunteers, and strategies. Her team developed a website containing a database of approximately 30 outreach organizations from the province and hosting outreach resources, a calendar of events, and professional development links for teachers. It also provided teachers with information for student-speakers to visit their classrooms and talk about science. Between 2009 and 2011, the website was a great success receiving nearly 4,500 hits.[42]

---

[41] Julita Vassileva, *Final Report, 2005–2011*, Regina, University of Saskatchewan, May 2012, p. 10–11; NSERC Chair for Women in Science and Engineering, *Regional CWSE Newsletter*, 2006, p. 4 and NSERC Chair for Women in Science and Engineering, *Regional CWSE Newsletter*, 2008, p. 4.

[42] Julita Vassileva, *Final Report, 2005–2011*, Regina, University of Saskatchewan, May 2012, p. 7; NSERC Chair for Women in Science and Engineering, *Regional CWSE Newsletter*, 2006, p. 4;

A second project in this area was the creation of a new website for *Canadian Women in Computer Science*. Julita's team took over the maintenance of this network from the Department of Computer Science at the University of British Columbia. Julita built a 200-entry database of female faculty members and instructors from Canadian computer science departments. The database was updated annually until Julita bumped into technical support problems in 2018. Finally, the Prairie Chair also served as the Ambassador for Canada in the Women's section of the Association for Computing Machinery (ACM-W), the main Computer Science professional organization. As the Canadian Ambassador from 2007 to 2011, she provided the ACM-W with information on programs, conferences, organizations, and people who were involved in solving the issues related to women in computing in our country.[43]

Besides her involvement in connecting women, Julita was engaged in two science outreach programs. As previously mentioned, she launched *myWISEmentor* in 2009 at the University of Saskatchewan to provide girls across the province with an opportunity to be mentored by women in science and engineering. Started in collaboration with the University of Saskatchewan College of Engineering Office of Outreach & Transition Programs, this initiative included an email mentoring program for girls between the ages of 11 to 18 in addition to access to a website containing several resources. Between 2009 and 2011, a total of 31 pairs of mentors and mentees were engaged in the process.[44]

Of all the programs that Julita had started during these six years, the *Science Ambassador Program* was her biggest achievement. The lack of exposure to exciting science education and role models partly explained why Indigenous female students were not choosing science or engineering programs at the University of Saskatchewan. Created in 2008, the *Science Ambassador Program* aimed to increase the interest of Indigenous youth in science and engineering. To do so, the program connected northern Indigenous communities with Julita's university by pairing senior undergraduate and graduate students in natural sciences, engineering, or health sciences with schools across northern Saskatchewan and Manitoba for four–six-week placements each spring. There they assisted the local science teacher by organizing materials, helping students, and finding hands-on, creative, and

NSERC Chair for Women in Science and Engineering, *Regional CWSE Newsletter*, 2008, p. 4 and NSERC Chair for Women in Science and Engineering, *Regional CWSE Newsletter*, 2010, p. 3.

[43] Julita Vassileva, Interview with Hannah Elizabeth Young, telephone interview, May 6, 2018; Julita Vassileva, *Final Report, 2005–2011*, Regina, University of Saskatchewan, May 2012, p. 8–9.

[44] Julita Vassileva, *Final Report, 2005–2011*, Regina, University of Saskatchewan, May 2012, p. 6.

culturally relevant activities to explain science concepts. In the first year, two ambassadors were sent to four communities—Muscoday and Beardy's & Okemasis First Nations in Saskatchewan in addition to six schools in The Pas and at the Opaskwayak Cree Nation in Manitoba. Soon the program spread into communities in all three Prairie provinces, growing to 11 ambassadors in 2011. In the first four years, 3,150 students participated in the *Science Ambassador Program.* Through questionnaires and qualitative feedback, Julita's team was able to see the positive change brought in Indigenous students' attitudes toward science and engineering and careers in these fields. But the best indication of its success is that every community that was involved in this program wanted to continue, and more schools expressed interest in joining each year. In 2017 alone, the program was offered in 12 communities, reaching more than 2,678 students, from which 94% were of First Nations or Métis ancestry.[45]

The West Coast Chair: Anne Condon (2003–2010)

With her nomination as the NSERC/General Motors Canada Chair for Women in Science and Engineering for British Columbia and the Yukon in 2004, Anne Condon was no stranger to this issue. A professor in the Department of Computer Science at

[45] Julita Vassileva, Interview with Hannah Elizabeth Young, telephone interview, May 6, 2018; Julita Vassileva, *Final Report, 2005–2011*, Regina, University of Saskatchewan, May 2012, p. 4–5; NSERC Chair for Women in Science and Engineering, *Regional CWSE Newsletter*, 2006, p. 4; NSERC Chair for Women in Science and Engineering, *Regional CWSE Newsletter*, 2008, p. 4; NSERC Chair for Women in Science and Engineering, *Regional CWSE Newsletter*, 2010, p. 3; University of Saskatchewan, *2017 Annual Report*, Saskatoon, Science Ambassador Program, 2017, p. 2, accessed April 6, 2019, https://artsandscience.usask.ca/scienceoutreach/documents/2017%20Final%20Report.pdf and James Shewaga, "Decade of Dedication for U of S Science Ambassadors", August 22, 2016, accessed April 6, 2019, https://news.usask.ca/articles/people/2016/decade-of-dedication-for-u-of-s-science-ambassadors.php

the University of British Columbia (UBC) since 1999 and a specialist in bioinformatics, biomolecular computation, and complexity theory, she has conducted many successful outreach programs to attract female students to science and engineering programs and careers. Indeed, in addition to her research on computational prediction of RNA and DNA secondary structure, with applications to the design of novel structures and gene synthesis, Anne had previously headed the *Distributed Mentor Project* as a professor at the University of Wisconsin. Through this program, approximately 70 female undergraduate students spent a summer conducting research under the supervision of a woman mentor. The third-party evaluation of the project showed that, compared to data from the National Center for Educational Statistics, a significantly higher proportion of the students involved in the project later entered graduate school. Because of its success and its thorough evaluation, the 2000 National Science Foundation (NSF) budget proposal to the US Congress featured it as one of four flagship projects. At the University of British Columbia, she created the *Focus on Women in Computer Science* committee, which organized activities to support female computer science students and recruit more women to the department. In addition to developing outreach programs at her host universities, Anne also served on the Computing Research Association's Committee on the Status of Women (CRA-W) between 1994 and 2007.[46]

Having spent over a decade in the United States, Anne's approach to her Chair was naturally shaped by that experience. Her involvement with the CRA-W proved to be more than significant since Anne appreciated the organization's general approach to the issue of women in science and engineering. She was particularly fond of the Distributed Mentor Project and the CRA-W network activities. Anne's work also bore the mark of the US NSF ADVANCE Program which had recently started in 2001, focusing on developing systemic approaches to increase the participation and advancement of women scientists and engineers in the academic profession and academic workplaces. A second influence on Anne's program was the unwillingness of her department head to reduce her teaching load in contrast to other Chairholders. This constraint led her to think about the best way to manage her teaching commitments with the Chair. One way to do this was to develop a course that would bring more women into computer science at UBC. Finally, as a woman computer scientist, Anne had the opportunity to witness the impact of the presence of numerous initiatives to support women in this field. Being part of a strong network such as the CRA-W or the Grace Hopper Conference was crucial for her own career. Thus, it was important for her to develop a regional network that

---

[46] "The Next Generation of Women Scientists", *UBC Reports*, vol. 50, no. 10 (November 4, 2004), accessed April 7, 2019, https://news.ubc.ca/2004/11/04/archive-ubcreports-2004-04nov04-generation/; "Interview with Anne Condon", *CRA Newsletter*, winter/spring 2009, accessed April 7, 2019, https://cra.org/cra-w/interview-with-anne-condon/; and "Anne Condon: 2014 Winner of the Grace Hopper Celebration Technical Leadership ABIE Award", 2014, accessed April 7, 2019, https://anitab.org/profiles/abie-award-winners/anne-condon/

would provide the same kind of opportunities for girls and women in British Columbia and Yukon.[47]

Like her predecessor Maria Klawe, Anne organized her Chair activities around her expertise in computer science. She designed a five-year project called the Jade Project, with reference to the official gemstone of British Columbia, jade, which is renowned for its beauty and strength. The objectives of the Jade Project were to "increase the participation and success of girls and women in the physical sciences and engineering, particularly computer science."[48] The program was divided into three core initiatives: outreach programs, course development, and research. With respect to outreach, the *Jade Bridges Programme* was created in 2004 to build institutional links between Anne's team and individuals at other colleges and universities in the West Coast region who were working on recruiting and supporting women in science and engineering. In its first two years, the program funded 31 projects targeting females from elementary to high school and undergraduate to graduate students at seven different universities and colleges in British Columbia. The activities included panels, workshops, and seminars for women undergraduate and graduate students, networking and mentoring programs connecting women role models in industry and academia, camps, and competitions to encourage high school girls to consider career options in science and engineering and focus groups to understand the needs of women science students. In its first two years, the program reached 530 women students from Grade 5 to postdoctoral fellows coming from different parts of the province. Feedback from surveys completed by project coordinators and departmental representatives were in general extremely positive. Through regular Jade Bridges Leaders Meetings, several project leaders exchanged experiences about their project and shared ideas for networking and collaboration, thus providing insight as to why the network was so useful.[49]

A second program developed by Anne's team, the *Canadian Distributed Mentor Project*, aimed to encourage undergraduate women in computer science and computer engineering to go to graduate school. The project funded outstanding women undergraduate computer science students to travel to a research institution for a summer of research under the supervision of a woman mentor. As the former coordinator of the CRA-W's *Distributed Mentor Project* (DMP), Anne knew firsthand

[47] Anne Condon, Interview with Hannah Elizabeth Young, telephone interview, March 12, 2018; Anne Condon, *The Jade Project: Interim Report*, Vancouver, University of British Columbia, June 2006, p. 1 and Yvonne Coady, Anne Condon, Anja Lanz, Andy Law, Michele Ng, Karen Parrish, and Beth Simon, "The Jade Bridges Project and Network", paper presented to the *12th CCWESTT Conference*, Guelph, May 29–31, 2008, accessed April 14, 2019, https://www.cs.ubc.ca/~condon/jadeproject/JADE-CCWESTT.pdf

[48] NSERC Chair for Women in Science and Engineering, *Regional CWSE Newsletter*, 2006, p. 5.

[49] Anne Condon, Interview with Hannah Elizabeth Young, telephone interview, March 12, 2018; "The Jade Project: Bridges Programme", unknown date, accessed April 14, 2019, https://www.cs.ubc.ca/~condon/jadeproject/bridges.html; Yvonne Coady, Anne Condon, Anja Lanz, Andy Law, Michele Ng, Karen Parrish, and Beth Simon, "The Jade Bridges Project and Network", paper presented to the *12th CCWESTT Conference*, Guelph, May 29–31, 2008, accessed April 14, 2019, https://www.cs.ubc.ca/~condon/jadeproject/JADE-CCWESTT.pdf and Anne Condon, *The Jade Project: Interim Report*, Vancouver, University of British Columbia, June 2006, p. 1–3.

how effective this kind of initiative could be. An external evaluation of the American program found that over 50% of surveyed DMP participants went on to graduate school as compared to about 3% of all women who earned undergraduate degrees in computer science in the United States. Building on this success, Anne launched the *Canadian Distributed Mentor Project* in 2005, a program that supported a total of 34 women in its four years of existence. As of 2010, at least 17 had pursued graduate school, and another eight planned to attend graduate school.[50]

The *Jade Treasures* was a third project aiming to provide resources and materials for high school students that dispel common myths about the field of computer science. The project also supported several programs that engaged girls in activities. Among these were the FIRST Lego League and the FIRST Robotics Competition which both aimed to expose students from Grades 6 to 12 to various aspects of engineering, computer science, and physics through robot building. To encourage more women students to participate in these competitions, Anne's team cooperated with the Vancouver School Board to provide funding and mentoring for schools which managed to form teams with at least 60% of their members being female. The *Jade Treasures* project also funded the *Girlsmarts Workshops* to introduce computer and engineering concepts and knowledge to girls in Grades 6 and 7 through a full-day workshop and hands-on activities, such as taking apart a computer, learning HTML and Java programming, cryptography, puzzles, and robot simulations.[51]

The teaching commitment constraints led Anne to develop the course *Connecting with Computer Science* in 2003, an interdisciplinary course introducing computer science fundamentals to university students through applications in the arts, psychology, linguistics, and biology. The design and the pedagogy of the course were inspired by feminist research. Thus, the course highlighted the contributions of both men and women in the field and supported different styles and approaches to programming. The programming assignments were also designed to enable students to

[50] "Anne Condon Wins A. Nico Habermann Award", *CRA-W Newsletter*, winter/spring 2010, accessed April 7, 2019, https://cra.org/cra-w/anne-condon-wins-a-nico-habermann-award/ and Anne Condon, *The Jade Project: Interim Report*, Vancouver, University of British Columbia, June 2006, p. 4–5.

[51] Anne Condon, *The Jade Project: Interim Report*, Vancouver, University of British Columbia, June 2006, p. 4 and Anne Condon, Interview with Ariane Zamin, telephone interview, 2017.

explore the connections between programming and creativity. Introducing all these changes in her course, Anne contributed to the wider effort of the University of British Columbia Department of Computer Science to attract more women. Because of the success of the course, the department decided to replace the original introductory computer science course with this new course and expanded the number of offerings. As of 2019, the course was still being offered on a regular basis.[52]

In regard to her research on women in Science, Technology, Engineering and Mathematics (STEM), Anne was involved in the Assessment Advisory Committee of the Faculty of Science at the University of British Columbia, formed in 2005. Supported by the Dean of the Faculty of Science and the Offices of the Provost and the Vice President, Research, the initial focus of the advisory committee was to assess the institutional environment and differences experienced by women and men faculty members in science. Following similar analysis on gender equity, recruitment, and retention that had been instigated at the Massachusetts Institute of Technology and other American elite universities, the study was soon expanded to comprise an assessment of the working climate. Led by Rachel Kuske, a professor at the Department of Mathematics at UBC, the study included a faculty survey, a department head questionnaire, and a collection of quantitative data from various administrative units across the campus. Focus groups were added to provide context for the data collected and to allow for more detailed responses to faculty members' concerns.[53]

Although faculty members overall reported a general sense of fair treatment and collegiality, results from the study showed that concerns about lack of professional support, issues of workplace equity, and career development needs were raised by women. Limited differences by gender were found in areas where clear policies were applied, such as salaries. In contrast, in areas such as time to promotion, amounts of retention funding, and awards, the lack of transparent and equitable procedures and policies led to significant gender differences. Several recommendations were put forward by the advisory committee and were well received by the Dean of Science, Simon Peacock. To help implement these recommendations, Simon asked Anne in 2007 to become the new Associate Dean of Faculty Affairs and Strategic Initiatives. The position was created as a result of the assessment with both a mandate and resources to support the wide range of concerns raised by Science faculty members, from recruitment and throughout their career at

[52] Anne Condon, *The Jade Project: Interim Report*, Vancouver, University of British Columbia, June 2006, p. 5–6 and Anne Condon, Interview with Hannah Elizabeth Young, telephone interview, March 12, 2018.

[53] Rachel Kuske, Carola Hibsch-Jetter, Simon Peacock, and Anne Condon, "Improving the Working Climate in the Faculty of Science at UBC", paper presented to the *12th CCWESTT Conference*, Guelph, May 29–31, 2008, accessed April 14, 2019, https://science.ubc.ca/sites/science.ubc.ca/files/faculty/diversity/CCWESTT.pdf and Rachel Kuske, Elizabeth Croft, Anne Condon, Grant Ingram, Janis McKenna, Wayne Maddison, and Michiel van de Panne, "An Assessement of the Working Climate for Science Faculty at the University of British Columbia (UBC)"—Executive Summary", Vancouver, University of British Columbia, May 2007, accessed April 7, 2019, https://apsc.ubc.ca/sites/default/files/legacy/2014-WCS-APSC-Report-FINAL.pdf

UBC. According to Anne, her work as associate dean was her best work because it gave her "a lot of scope to work on faculty related projects."[54] Since the implementation of several recommendations from the working climate report, the representation of women among full professors across the Faculty of Science has increased drastically from 3% in 2000 and 8% in 2008 to reach 18% in 2014.[55]

***

During their time as Chairs for Women in Science and Engineering, the five women of the second generation of CWSE Chairs were able to build a strong National Network. With NSERC's decision to provide funds for the National Network, Cecilia, Nadia, Valerie, Julita, and Anne interacted to a much greater extent than the previous generation of Chairholders. Since 2006, the five regional programs have shared information and resources; established regular face-to-face meetings; given a greater visibility and impact at a national level; and provided research opportunities that support the common objectives of the program. This last component proved to be a notable aspect of their collaboration. Undeniably, bringing together a research network that included the regional Chairs as a nucleus in addition to researchers from social science and humanities was truly innovative. The very complex and multidimensional problem of the underrepresentation of women in science and engineering required an interdisciplinary team with a connection to social sciences and humanities. By developing such research at the national level, the second generation of Chairs was beginning to be truly a team working toward the same goal, instead of a group of individuals working independently on similar topics. The network also facilitated liaisons with national organizations such as CCWESTT and Engineers Canada as well as international bodies like INWES.

In addition to the creation of a formal National Network, the second generation of regional Chairholders established lasting regional and national programs. For example, Cecilia Moloney developed the *MetaKettle Project* to change education to make engineers more socially aware as a result of her involvement as Atlantic Chair, a program that is still running today. The Ontario Chair developed the program *Go Eng Girl/GÉNiales, les filles* that has spread throughout Canada and is still in operation almost 20 years after its inception. The Quebec Chair partnered with the

[54] Anne Condon, Interview with Hannah Elizabeth Young, telephone interview, March 12, 2018.

[55] "Work climate, diversity and equity", date unknown, accessed April 14, 2019, https://science.ubc.ca/faculty/diversity; "Interview with Anne Condon", *CRA Newsletter*, winter/spring 2009, accessed April 7, 2019, https://cra.org/cra-w/interview-with-anne-condon/; Rachel Kuske, Carola Hibsch-Jetter, Simon Peacock, and Anne Condon, "Improving the Working Climate in the Faculty of Science at UBC", paper presented to the *12th CCWESTT Conference*, Guelph, May 29–31, 2008, accessed April 14, 2019, https://science.ubc.ca/sites/science.ubc.ca/files/faculty/diversity/CCWESTT.pdf and Rachel Kuske, Elizabeth Croft, Anne Condon, Grant Ingram, Janis McKenna, Wayne Maddison, and Michiel van de Panne, "An Assessement of the Working Climate for Science Faculty at the University of British Columbia (UBC)"—Executive Summary", Vancouver, University of British Columbia, May 2007, accessed April 7, 2019, https://apsc.ubc.ca/sites/default/files/legacy/2014-WCS-APSC-Report-FINAL.pdf

Marianne-Mareschal Chair to set up the event *Les filles et les sciences: un duo électrisant!* in Quebec City for the first time in 2008, which is still a success 20 years later. The Prairie Chair created the *Science Ambassador Program* reaching over 3,000 Indigenous students in its first four years of operation, which is still in operation. Finally, the *Connecting with Computer Science* course established by the West Coast Chair in 2003 is still taught in Anne's department. These five examples are but just a fraction of all the initiatives developed by the second generation of regional Chairs. As in the previous generation, some of the Chairholders moved on to other projects, while others continued to work as advocates for women in science and engineering. Together, the five regional Chairs were able to further increase the awareness about the participation of women in these disciplines across Canada. Building upon their predecessors' accomplishments and ideas, and as their predecessors did, they elevated and evolved the interventions and research to address the multifaceted problem of underrepresentation of women in science and engineering, setting the scene for the next generation of Chairholders to come.[56]

[56] Lesley Shannon and Catherine Mavriplis, "Canada's Chairs for Women in Science and Engineering Program: 20 Years of Strategies to Increase the Participation of Women", paper presented to the *9th European Conference on Gender Equality in Higher Education*, Paris, France, September 13, 2016.

Chapter 4
The Third and Fourth Generations of Regional Chairs (2011–2021)

Starting in 2010, a new generation of women gradually took up the role of regional Chairs: Dr. Elizabeth Croft, a professor in the Department of Mechanical Engineering at the University of British Columbia, was sponsored by multiple donors from 2010 to 2015 as the Chair for the British Columbia and Yukon region. In 2011, three new Chairs were appointed: Dr. Tamara Franz-Odendaal, a professor in the Department of Biology at Mount Saint Vincent University, held the Atlantic region Chair, whose industry sponsors included Eastlink (2011–2016) and, since 2017, Lockheed Martin Canada; Dr. Catherine Mavriplis, professor in the Department of Mechanical Engineering at the University of Ottawa, as the Ontario Chair, was sponsored first by Pratt & Whitney Canada for the 2011–2016 term and, from 2017 to 2021, by the Leacross Foundation, CAE and Lockheed Martin Canada; and Dr. Annemieke Farenhorst, a professor in the Department of Soil Science at the University of Manitoba, led the Prairie region Chair, which was funded by multiple donors from 2011 to 2020. In 2015, two new appointments were made: Dr. Ève Langelier, a professor in the Department of Mechanical Engineering at the Université de Sherbrooke, took on the Quebec Chair which was funded by multiple donors; and Dr. Lesley Shannon, a professor in the School of Engineering Science at Simon Fraser University, effectively started a fourth generation of Chairs in the British Columbia and Yukon region. Her Chair was funded first by Simba Technologies Inc. and then by Orbital Technologies Inc. (Fig. 4.1).

As for the second generation, the new Chairholders could count on their predecessors to act as mentors and offer advice and wisdom. The overlap between generations proved helpful in many aspects of the Chair mandate. First, Valerie Davidson organized a meeting of the Chairs in the fall of 2011 in Guelph, which included Natural Sciences and Engineering Research Council of Canada (NSERC) program managers and media training by NSERC. Elizabeth Croft, who had already been Chair for more than a year, and her program manager, Jennifer Pelletier, assisted in the transition of Tamara, Catherine, and Annemieke by sharing organizational best

C. D'Amours et al., *The Chairs for Women in Science and Engineering Program*, Women in Engineering and Science, https://doi.org/10.1007/978-3-031-54085-1_4

Fig. 4.1 First meeting of the third generation of Chairs in 2011 in Guelph. Clockwise from 12 o'clock: Nadia Ghazzali, Catherine Mavriplis, NSERC Program Manager Enikö Megyeri-Lawless, Tamara Franz-Odendaal, Elizabeth Croft, Annemieke Farenhorst, and Valerie Davidson

practices and strategies. The National Network kept the group in close contact during the crucial start-up phases of the new Chairs. Further, Lesley Shannon had the opportunity to settle in smoothly in her new role as Elizabeth Croft's tenure overlapped with hers for six months in 2015. For example, Lesley was able to build relationships in the Yukon while accompanying Elizabeth on her last trip there. She also was in contact with Elizabeth's program manager, Jennifer Pelletier, who drafted a job description for the same position with her, and the two collaborated for staff training, thus enabling Lesley to efficiently go through the process of hiring a key figure of her new team. The continuity between the two Chairholders was most helpful for Lesley's transition to her new role. Yet, not everyone had the same opportunity as Lesley. In the Atlantic and Quebec regions, there was no overlap, and in fact there were gaps. Fortunately for Tamara, and later Ève, the very existence of the National Network and the fruitful collaborations developed with Chairs from across Canada proved to be very beneficial. Ève and Catherine also had former Chairs on their advisory boards, Claire Deschênes for Ève, and Mary Williams for Catherine. Thus, the transition between generations was significantly easier than the previous one.[1]

[1] Lesley Shannon, Interview with Hannah Elizabeth Young, telephone interview, May 31; Tamara Franz-Odendaal, Interview with Hannah Elizabeth Young, telephone interview, June 1, 2018; Tamara Franz-Odendaal, Interview with Ariane Zamin, telephone interview, January 18, 2018; and Elizabeth Croft, *Second 24-Month Progress Report (Months 24–48)*, Vancouver, University of British Columbia, 2014, p. 6.

Strengthening the National Network

Since its official inception in 2006, the National Network of NSERC Chairs for Women in Science and Engineering developed mechanisms to facilitate interactions among the five Chairs. Between 2011 and 2019, the regional Chairs worked together on several national initiatives. Following the practice of the previous generations, one Chairholder acted as coordinator of the National Network. From 2010 to 2015, Elizabeth led the group and managed the budget and, between 2015 and 2019, Catherine carried the torch. In 2020, Tamara took over as network manager. As a group, the network enhanced communication among the five Chairs through regular face-to-face meetings in various cities across Canada as well as teleconferences. Over time, the contact increased to a monthly meeting, with one to two of these being all-day in-person working sessions, typically at the Canadian Coalition for Women in Science, Engineering, Technology and the Trades (CCWESTT) conference or a similar venue. The increased contact also amplified the effectiveness of the five regional programs through shared information and resources. For example, the Chairs agreed to use the National Network funds to sponsor academic advancement workshops for assistant professors targeting tenure, and associate professors aspiring a promotion to full professor, following models developed by Catherine. These workshops focused on "skills, strategies, and 'insider information' necessary to obtain and succeed in a tenure-track position in science and engineering and advance to other positions of leadership."[2] Several different workshops were offered to 700 women professors or those applying to become professors to look ahead to promotion and get advice and support on their career path: *Step-Up*, *FORWARD to Professorship*, *Promotion to Professorship*, *Women in Science and Engineering Advancement*, *Tenure and Beyond*, *and Faire progresser votre carrière universitaire*. Between 2014 and 2022, the National Network supported 11 workshops throughout the country: in Halifax (2014), Ottawa (2015), Vancouver (2015), Calgary (2016), Hamilton (2017), Saint-John (2017), Burnaby (2018), Toronto (2019), Dartmouth (2020), Atlantic region (online in 2020), and Sherbrooke (2022).[3] In a similar effort to distribute resources evenly across the country, the National Network established the National Network Conference Grant in 2017, later renamed for the late Margaret-Ann Armour, to provide support to the many conferences organized by stakeholder groups for the advancement of women in various science and engineering fields. From 2017 to 2019, the network received 47 applications and

---

[2] Chairs for Women in Science and Engineering National Network, *Coast to Coast update 2017*, Ottawa, Réseau National Network, 2017, p. 1.

[3] Chairs for Women in Science and Engineering National Network, *Coast to Coast update 2017*, Ottawa, Réseau National Network, 2017, p. 1; Chairs for Women in Science and Engineering National Network, *Coast to Coast update 2018*, Ottawa, Réseau National Network, 2018, p. 1; Catherine Mavriplis, *48-month Report*, Ottawa, University of Ottawa, August 2015, p. 1 and CWSE-ON, "National Network", date unknown, accessed August 7, 2019, http://www.cwse-on.ca/cwse/national-network

funded 19 requests covering 24 disciplines, from mining to ecology, and 30 regions of Canada, including Fredericton, Guelph, Saskatoon, and Edmonton.

Through the National Network, the five Chairholders also increased the visibility and impact of Chair for Women in Science and Engineering (CWSE) activities at the national and international levels. For instance, all five were particularly active at the two international Gender Summits held in North America, in Washington, D.C. in 2013 and in Montreal in 2017. These events aimed to connect a large swath of relevant stakeholders to encourage commitment to progress and engagement in implementing equity, diversity, and inclusion actions of biological sex and gender considerations in research and innovation fields. At both events, which gathered more than 600 participants from academia, government, and industry, the Chairholders presented workshops and panels to introduce the Chairs program to their international counterparts. At the Montreal event, the Chairs, especially Ève, were deeply involved in the organization of the event. At the national level, the Chairs participated in Equity, Diversity, and Inclusion (EDI) consultations for the Tri-Council representing Canada's national funding agencies in Medicine and Health Sciences (Canadian Institutes of Health Research, CIHR), Science and Engineering (NSERC), and the Humanities and Social Sciences (Social Sciences and Humanities Research Council of Canada, SSHRC), as well as the EDI task force initiated by the Fonds de recherche du Québec, the Québec province research funding agency, in 2017.[4] The Chairs established a standing presence at each biennial national CCWESTT conference over the years, organizing talks, panels, and activities, and in some cases the whole conference (in 2012, 2016, and 2022), publishing a Coast-to-Coast Update brochure of their activities for widespread distribution at the event.

The Chairs contributed to the establishment of the *Canadian Archive of Women in STEM* at the University of Ottawa in partnership with Library and Archives Canada in June 2018. The project began as a workshop, the *Creating the Memories and Celebrating the Legacy of the Bold and the Brave: Building the Archives of Women Scientists and Engineers in Canada* held in September 2014. The workshop gathered 45 archivists, historians, women scientists and engineers, science journalists, and policymakers to discuss best practices to collect, preserve, develop, and share records belonging to women in Canadian Science, Technology, Engineering and Mathematics (STEM). Organized in partnership with previous Chairholders Monique Frize and Claire Deschênes, and their historian colleague Ruby Heap, the goal of the workshop was "the creation of a taskforce mandated to work toward the creation of a national archive that would help build the history of women in Canadian

[4] NSERC, "NSERC at the Gender Summit 3—North America", *Contact Newsletter*, vol. 38, no 1 (Winter 2014), accessed August 8, 2019, http://www.nserc-crsng.gc.ca/Media-Media/Newsletter-Bulletin/v38_n1/v38_n1_eng.asp#a9; Catherine Mavriplis, *24-month Progress Report*, Ottawa, University of Ottawa, September 2013, p. 5; Elizabeth Croft, *Second 24-Month Progress Report (Months 24–48)*, Vancouver, University of British Columbia, 2014, p. 13 and Catherine Mavriplis, *24-month Report*, Ottawa, University of Ottawa, June 2019, p. 6–7.

STEM and celebrate their largely unknown legacy."[5] With financial support from several organizations and in the presence of the federal Minister of Science, the Honorable Kirsty Duncan, and the Chief Science Advisor, Mona Nemer, the archive was officially launched on June 19, 2018, at the University of Ottawa where the records were to be preserved.[6]

The National Network worked together on the development of a performance management strategy (PMS) for the CWSE program and other activities supporting the participation of women and girls in STEM. In 2013, NSERC program managers had asked the Chairs for Women and the Chairs for Design Engineering to come up with a common set of indicators for their activities. From 2014 onward, the National Network reported impact and outcomes based on 12 indicators (predominantly focusing on activities, outputs, and reach) to NSERC and through their Coast-to-Coast Update publications. While these indicators captured important numerical data for the impact of the Chairs program, long-term outcomes and more qualitative data were still difficult to distill on a national scale given the varied nature of the five Chair programs. Furthermore, none of the Chairs was using a performance measurement framework to evaluate the vast array of indirect delivery mechanisms they were involved in. Thus, the five regional Chairs agreed in 2017 to work together on developing a PMS to provide better measures of the outcomes and impacts of their work. Collaborating with Jane Whynot, a consultant in evaluation and performance measurement, the National Network developed a PMS based on the COM-B theory of change model, which postulates that behavior change (B) only occurs if capabilities (C), opportunities (O), and motivation (M) are present. An article on the development of the framework was published in the *Canadian Journal of Program Evaluation* in 2019, and a draft version of their PMS was enthusiastically received by NSERC to consider in their own evaluation of the Chairs program which took place in 2021. This collaboration ensured that each Chair would eventually use the same framework, which would then allow the Chairs (and NSERC) to measure the national impact of the NSERC CWSE program using this framework.[7]

The National Network was also the catalyst behind research activities that support the common objectives of the CWSE Program. Among several other research collaborations, there were three important projects. First, lacking comprehensive data for the statistics of women's participation in science, the National Network

[5] INWES Education and Research Institute, "Research Committee, Report 2017–2018", News, 2018, date accessed August 8, 2019, https://inwes-eri.org/2019/03/19/research-committee-report/

[6] Tamara Franz-Odendaal, *48-month Progress Report*, Halifax, Mount Saint Vincent University, February 2016, p. 8–9 and INWES Education and Research Institute, "Research Committee, Report 2017–2018", News, 2018, date accessed August 8, 2019, https://inwes-eri.org/2019/03/19/research-committee-report/

[7] Annemieke Farenhorst, *2019 Progress Report*, Winnipeg, University of Manitoba, 2019, p. 6–7; Catherine Mavriplis, *24-month Report*, Ottawa, University of Ottawa, June 2019, p. 6–7 and Jane Whynot, Catherine Mavriplis, Annemieke Farenhorst, Ève Langelier, Tamara Franz-Odendaal, and Lesley Shannon, "Knitting Theory in STEM Performance Stories: Experiences in Developing a Performance Framework", *Canadian Journal of Program Evaluation*, vol. 33, no 3 (2019), p. 354–374.

presented the report *Analysis of the distribution of gender in STEM fields in Canada* as a national resource in 2018. The document offered a statistical overview of the distribution of gender in science and engineering fields in Canada in terms of undergraduate and graduate degrees awarded by sex, major field of study by sex, and wages and employment in STEM fields by sex. Based on the 2011 National Household Survey, the Labour Force Survey 2001–2016, and the National Graduates Survey 2000–2009, the report was a resource to guide students, professionals, policymakers, and society as a whole to inform their thinking and approach to diversity and inclusion in these domains.[8] Throughout the National Network projects, several resources, including this report, were translated into French. Second, through the National Network, a university climate survey was undertaken in three of the five Chair regions, providing a national picture of academic culture in Canada. The results have been published in academic journal papers. Third, the participation of the Chairs in the 2014 *Creating the Memories and Celebrating the Legacy of the Bold and the Brave: Building the Archives of Women Scientists and Engineers in Canada* workshop provided the impetus for the Chairs to commit to documenting the work of the Chairs for a lasting record. The result was the drafting and publication of the book you are now reading.

Throughout all these collaborations, the Chairholders strengthened the National Network while working to increase the participation of women in science and engineering throughout the country. If the five regional Chairs aspired to achieve the same goals, they set their own course of action according to their personal and professional interests, their strengths, and their regional geographical and political settings. Although each Chair's agenda was very different in content, audience, and scope, they were all successful in their attempts to promote science and engineering to women and create greater gender inclusivity in these fields.

The Atlantic Chair: Tamara Franz-Odendaal (2011–2023)

[8] National Network of Chairs for Women in Science and Engineering (CWSE), *Analysis of the Distribution of Gender in STEM Fields in Canada*, Ottawa, National Network CWSE, 2018, 71 p. and CWSE-ON, "National Network", date unknown, accessed August 7, 2019, http://www.cwse-on.ca/cwse/national-network

When she was nominated as the NSERC Chair for Women in Science and Engineering for the Atlantic Region (WISEAtlantic) in 2011, Dr. Tamara Franz-Odendaal had been a professor in the Department of Biology at Mount Saint Vincent University for five years. A researcher with a strong background in evolutionary-developmental biology, Tamara's research interests focused on comparative vertebrate development and evolution, with particular attention to the neural crest–derived craniofacial skeleton. Her interest in the Chair position came from two places. Her Associate Vice President of Research at the time first pointed out the CWSE position in 2010 and asked her if she was interested. Since Tamara had done a considerable amount of science outreach in South Africa before moving to Canada, she knew that she would enjoy it. Furthermore, she had two women supervisors (both mothers) during her PhD studies, and as a mother herself, serving as a role model to other women scientists appealed to her. After talking with the former Prairie Chair, Julita Vassileva, Tamara was convinced that this initiative was worth her time.[9]

Tamara's work as the Atlantic Chair was impacted by established presences in the region but also the distance from Newfoundland, where the two previous Atlantic Chairs had been. First, the presence of several strong organizations interested in STEM, such as the Newfoundland and Labrador Women in Science and Engineering (WISE) chapter and Techsploration, both organizations that do important work with girls in the region, gave Tamara the possibility of building strong partnerships right from the start. This made even more sense when taking into account the Atlantic region's large geography, the second major influence on Tamara's activities. Having to cover a territory that extended through several provinces, she quickly understood that she had to adapt to this reality and made her own resources completely accessible to the entire region. The location of the Atlantic Chair before her arrival also had a strong influence on her work. Since the last two Chairholders were based in Newfoundland, it was difficult for them to have a big impact outside their own province. In addition to this, there was a gap of two years between Cecilia Moloney's tenure and Tamara's, leaving the latter with little to build on. This lack of continuity led Tamara to build her own program from scratch, as did many other Chairholders.[10]

The central activities of the Atlantic Chair were divided into four general areas: youth engagement, raising awareness and engaging others, providing support for women in STEM, and research. In terms of youth engagement, the *Girls Get WISE Science Camp* was one of the Chair's signature events. The camp, which had a capacity of 25–35 students, targeted girls in Grades 7 and 8, typically

[9] Tamara Franz-Odendaal, Interview with Hannah Elizabeth Young, telephone interview, June 1, 2018; NSERC, "Chairholder Profile: Tamara Franz-Odendaal", May 20, 2014, accessed July 24, 2019, http://www.nserc-crsng.gc.ca/Chairholders-TitulairesDeChaire/Chairholder-Titulaire_eng.asp?pid=764 and Mount Saint Vincent University, "Tamara A. Franz-Odendaal", date unknown, accessed July 24, 2019, https://www.msvu.ca/en/home/programsdepartments/bachelorofscience/biology/facultyprofiles/TamaraFranz-Odendaal/default.aspx

[10] Tamara Franz-Odendaal, Interview with Hannah Elizabeth Young, telephone interview, June 1, 2018, and Tamara Franz-Odendaal, *48-month Progress Report*, Halifax, Mount Saint Vincent University, February 2016, p. 11.

12–14-year-olds. This five-day summer camp focused on science-related workshops, games, demonstrations, and lab activities. Each camp featured the use of a common pet-store fish and model organism in developmental biology, the zebrafish, as a teaching tool and included other STEM-related hands-on activities beyond biology (e.g., design challenges, coding, etc.). Each camp included informal discussions with women working in a variety of STEM fields. The camp was such a success that Tamara received multiple requests from parents to create week-long day camps for younger and older girls. As a result, leveraging NSERC PromoScience funding, she launched a second week-long day camp for girls aged 15–16 years old in 2017 in addition to the first one. During this first week of activity, the senior girls explored science through presentations and demonstrations and had the opportunity to meet the federal Minister of Science, the Honorable Kirsty Duncan.[11]

Tamara's team also developed the one-day *Girls Get WISE Science Retreat* for girls in Grades 7–12. The event exposed girls to hands-on activities in STEM fields such as chemistry and programming robots. It incorporated role model sessions, in which girls spent time with professional women talking about their careers and how they enjoyed their jobs. Tamara's team also created a "how-to" guide on organizing and running the *Girls Get WISE Science Retreat*. This led the Atlantic Chair to partner with the Halifax Regional School Board—African Nova Scotian girls, Girl Guides Nova Scotia, the Association of Professional Engineers of the Province of Prince Edward Island (Engineers PEI), the Dalhousie Agricultural Campus WISE group, Acadia University, the University of Prince Edward Island, and Cape Breton University. As a result, the retreat was offered 20 times between 2012 and 2018. During this same six-year span, the Atlantic Chair was able to engage 557 girls and exposed them to 120 female role models in STEM through the *Girls Get WISE Science Camp* and the *Girls Get WISE Science Retreat* alone. Evaluations from these events indicated that they met the expectations of 96% of participants. More significantly, 90% of them said that they would consider a career in STEM because these activities opened their eyes to more career opportunities.[12] In addition to this direct engagement with girls, Tamara's program developed a series of very popular role model videos (with their partner Techsploration), a series of career posters for teachers to display in their classrooms, and a *Career Booklet* series featuring women in science, engineering, and computer science. These booklets showcase the diversity of women working in these fields. The final booklet in the series features Indigenous women in STEM fields. The booklets have been translated into French.

[11] "Girls Get WISE Science Camp August 13–17, 2012", *WISEatlantic Newsletters*, issue 2, September 2012, p. 1; WISEatlantic website, "Girls Get WISE Science Summer Camp", date unknown, accessed July 25, 2019, http://wiseatlantic.ca/ggwcamps/; Tamara Franz-Odendaal, *48-month Progress Report*, Halifax, Mount Saint Vincent University, February 2016, p. 7 and "NEW Girls Get WISE Science Camp for Girls Ages 15 & 16 Launched", *WISEatlantic Newsletters*, issue 17, Fall 2017, p. 3.

[12] Tamara Franz-Odendaal, *48-month Progress Report*, Halifax, Mount Saint Vincent University, February 2016, p. 7; WISEatlantic website, "Girls Get WISE Science Retreats", date unknown, accessed July 25, 2019, http://wiseatlantic.ca/girls-get-wise-science-retreats/

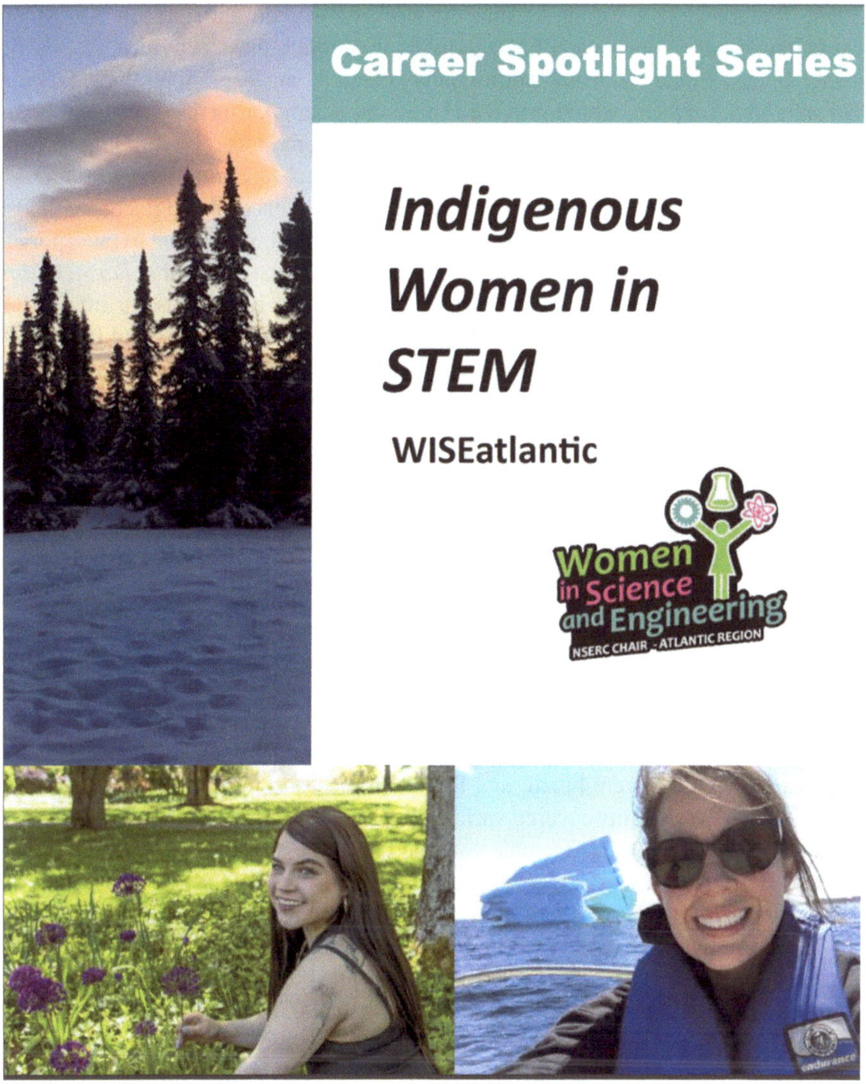

To raise awareness about women in STEM in Atlantic Canada and connect with other organizations with the same goal, Tamara developed a number of regional initiatives. First, she organized quarterly *Regional connection nodes* teleconferences with members of organizations throughout each of the Atlantic provinces that support women in STEM and encourage girls to pursue careers in these fields. The purpose of these teleconferences was to obtain information on the various activities and events of organizations with shared goals and to learn how to better support each other. They provided important connections in the region that previous Chairs had achieved with an advisory board. The Atlantic Chair had direct participation with 29 communities in the region and was involved with 42 organizations. As of

2016, Tamara built a strong Atlantic network of women and girls in STEM by exposing and engaging over 12,500 individuals through these activities. Along the same lines, Tamara's team additionally created in 2017 the *WISEAtlantic Partnerships Program*, which provided small one-year sponsorships for local organizations. It was an opportunity for community groups to receive funding to help establish or expand programs that promote the outreach, recruitment, and retention of girls, young women, and industry professionals in science and engineering fields. To ensure the sustainability of these programs, applicants were asked to incorporate plans for continuation after funding ended. In the first year, the *Partnerships Program* had six applicants, and three successfully received funds for innovative programs in 2018. The 2019 competition saw many more applicants and four organizations or individuals received funding up to $3,000 for their new project.[13]

In support of women professionals in STEM, the Chair hosted three *Atlantic ConnecTions Conferences* in 2015, 2017, and 2019. The idea to launch these conferences stemmed from Tamara's discussions with the other NSERC Chairholders. After hearing the success of regional conferences such as the *Creating Connections* conference series held in British Columbia by Elizabeth Croft, the British Columbia and Yukon Chair, and noticing that very few participants from Atlantic Canada attended the 2014 CCWESTT conference in Regina (despite earlier success when it was held in Halifax, Nova Scotia, in 2012), Tamara recognized the need for an Atlantic regional conference. The first *Atlantic ConnecTions Conference* was held in Sackville, New Brunswick, in June 2015, and gathered 75 participants from businesses, universities and colleges, not-for-profit organizations, and government. The idea was to help "women to connect with each other and leave the conference knowing that their voices were heard, and that their successes were celebrated."[14] Over two days, the participants covered various topics ranging from finding family-work-life balance to the difference female voices could make and the changing workplace culture. The overall conference expectations were met or exceeded by 95% of respondents, and 95% of them said that they would be interested in attending another conference in 2017. These evaluations set the stage for the second conference held in Halifax two years later. The focus of the 2017 event was on understanding the unique challenges in Atlantic Canada and on taking actions that would serve women in the region. Over 100 participants from across STEM sectors throughout the region responded to the invitation. The three-day program included keynote speakers, several professional workshops on various topics such as "Networking & Professional Presence," two parallel sessions on four different streams (women in engineering, women in trades, women in academia, and women in technology), and

[13] "Regional Nodes… Stay Connected", *WISEatlantic coas*, issue 11, Fall 2015, p. 3; Tamara Franz-Odendaal, *48-month Progress Report*, Halifax, Mount Saint Vincent University, February 2016, p. 1–3; WISEatlantic website, "News", date unknown, accessed July 25, 2019, http://wiseatlantic.ca/news/# and WISEatlantic website, "WISEatlantic Partnerships Program", date unknown, accessed July 25, 2019, http://wiseatlantic.ca/partnerships-program/

[14] Tamara Franz-Odendaal, *Atlantic ConnecTions Conference Report*, Halifax, Mount Saint Vincent University, July 2015, p. 3, accessed July 25, 2019, http://atlanticconnections.ca/wp-content/uploads/2015/07/15-07-15-Report-ACT.pdf

a Conversation Café. Again, the feedback was overwhelmingly positive and gave the impulse for the 2019 Conference in Charlottetown, Prince Edward Island.[15] Tamara's Chair also sponsored numerous workshops from the WinSETT Leadership Program series that have been delivered to women (students, apprentices, and professionals in the workforce) throughout the Atlantic region.

In terms of her research on women in STEM, two projects clearly stand out. The first one investigated the decision-making strategies used by girls while they grappled with their career decisions. Tamara's team developed an online survey that questioned students about subject preferences and competencies, STEM engagement, career influencers, STEM career awareness, and knowledge of engineering. Almost 1,000 students in Grades 7–9 from 21 schools across three Maritime Provinces answered the survey. The results showed that high engagement in STEM activities was the primary positive influencer directing students to a STEM career. The study also indicated that the younger students (in Grade 7) did not grasp the importance of science and math requirements for future careers in these fields. In phase two, a sample of 1,448 students in Grades 7 and 9 was drawn from Atlantic Canada schools to explore students' knowledge of science and mathematics requirements for STEM careers, their mathematics self-efficacy, their future career interests, their preferences for particular career activities, and their likelihood to pursue a career in these fields. Tamara's team concluded that, although older students had more knowledge about the requirements for these careers, this knowledge was lacking overall. Additionally, students with a higher mathematics self-efficacy and STEM career knowledge were more likely to pursue these careers as were students with greater interest in technical and scientific skills.[16] A more recent gender-based analysis of this data and of the girls who participated in the summer camps has revealed some interesting gender differences.

The second research project focused on faculty workplace climate. Partnering with Annemieke Farenhorst, the Prairie Chair, Tamara's team surveyed 395 faculty members at seven Atlantic Canadian universities in natural sciences, engineering, and computer science departments. The study showed that Atlantic faculty were

---

[15] Tamara Franz-Odendaal, *Atlantic ConnecTions Conference Report*, Halifax, Mount Saint Vincent University, July 2015, 81 p., accessed July 25, 2019, http://atlanticconnections.ca/wp-content/uploads/2015/07/15-07-15-Report-ACT.pdf; Tamara Franz-Odendaal, *Atlantic ConnecTions Conference Report*, Halifax, Mount Saint Vincent University, July 2015, 43 p., accessed July 25, 2019, http://atlanticconnections.ca/wp-content/uploads/2014/12/2017-Report-ACT-June-14-FINAL-.pdf and Chairs for Women in Science and Engineering National Network, *Coast to Coast update 2018*, Ottawa, Réseau National Network, 2018, p. 11.

[16] Tamara Franz-Odendaal, Karen Blotnicky, Frederick French, and Phillip Joy, "Experiences and Perceptions of STEM Subjects, Careers, and Engagement in STEM Activities Among Middle School Students in the Maritime Provinces", *Canadian Journal of Science, Mathematics and Technology Education*, vol. 16, no. 2 (2016), p. 153–168; Karen Blotnicky, Tamara Franz-Odendaal, Frederick French, and Phillip Joy, "A Study of the Correlation between STEM Career Knowledge, Mathematics Self-Efficacy, Career Interests, and Career Activities on the likelihood of Pursuing a STEM Career Among Middle School Students", *International Journal of STEM Education*, vol. 5, no. 1 (2018), 22 p. and Tamara Franz-Odendaal, *48-month Progress Report*, Halifax, Mount Saint Vincent University, February 2016, p. 6.

generally "more optimistic about the climate of their own departments/units relative to the broader university climate." However, the results indicated that women faculty members had more negative day-to-day workplace experiences such as harassment and discrimination and had higher levels of job-related stress. The study concluded that, nevertheless, both men and women faculty members were "generally satisfied with their own career progress" and most planned "to continue working at their current university."[17]

The Quebec Chair: Ève Langelier (2015–)

At the time of her appointment as the NSERC Chair for Women in Science and Engineering for the Quebec Region in 2015 (CFSG), Ève Langelier had been a professor in the Department of Mechanical Engineering at Université de Sherbrooke for 11 years. Being interested in human biomechanics and the mechanobiology of soft connective tissues, she studies the effects of forces that act on living tissues, such as tendons, and how they adapt to these mechanical forces. Her work focuses on the prevention and cure of tendon injuries as well as the design of adapted sports equipment.[18] In addition to her work in the field of bioengineering, Ève played an active role in organizing a variety of activities for the promotion, retention, and professional development of women in science and engineering. Her time as the regional representative of Claire Deschênes's Quebec Chair from 2004 to 2006 was the main contributing factor to her interest in these issues. As part of her mandate as regional representative, she also served on the Board of Directors of AFFESTIM. In the following years, she continued to run several initiatives to encourage girls and women to consider careers in science. In 2007, she started the activity *Les filles et*

[17] Jennifer Dengate, Annemieke Farenhorst, Tracey Peter, Tamara Franz-Odendaal, Danielle Saj, Sally Marchand, and Mahalia Lepage, *Faculty Workplace Climate Survey Results—Atlantic Canada Report*, Halifax, NSERC Chairs for Women in Science and Engineering, 2018, p. 3–5.

[18] "Ève Langelier—Engineering: It's not just a guy thing", May 22, 2015, accessed May 12, 2019, http://www.nserc-crsng.gc.ca/Women-Femmes/Langelier-Langelier_eng.asp and "Chairholder Profile: Ève Langelier", May 20, 2014, accessed May 12, 2019, http://www.nserc-crsng.gc.ca/Chairholders-TitulairesDeChaire/Chairholder-Titulaire_eng.asp?pid=912

les sciences, un duo électrisant! in the Eastern Townships. She also helped to create a toolkit for teaching the design approach used by engineers and conducted coaching activities for women and men students at her host university.[19]

Ève's work as the Quebec Chair was influenced by government policies in the province. First, as was the case with the former Quebec Chairs, Claire Deschênes and Nadia Ghazzali, the Quebec government's continuing concern with the skilled labour force proved to be beneficial for Ève's work. The provincial government was instrumental in the establishment of her Chair. Indeed, she had trouble at the beginning to secure funding from industrial partners. Thanks partly to the financial support of the Fonds de la recherche du Québec—Nature et technologies (Quebec Research Fund—Nature and Technologies or FRQNT), the Secrétariat à la condition féminine (the Status of Women Office) and the Ministry of Economy, Science and Innovation, she was able to present matching funds for NSERC funding. The implementation of the Quebec educational reform in the mid-2000s also influenced her Chair program. With engineering content now being included in the technological world ("l'univers technologique"), the formal name given to secondary school science curricula, Ève was well poised to contribute to teacher training as an engineer herself since several government reports and research conducted in the province highlighted the problems encountered by teachers of the new curricula. A majority of secondary school teachers were initially trained in science and not in engineering and thus did not receive the necessary training in engineering during their undergraduate studies. This problem would give her a starting point for her program.[20]

In addition to her work in bioengineering, the Quebec Chair also focused her time on public awareness, outreach activities, and research. With respect to public awareness, her first major initiative was the generation of accessible resources. The Chair produced and directed two series of videos *Changer le monde au féminin* et *D'une femme à une autre* which were aimed primarily at young women in secondary school and Collège d'enseignement general et professionnel (CEGEP). With the help of 23 female role models, these videos demonstrated that science and engineering can also be people-oriented despite misconceptions present in Quebec society. They showed how science and engineering jobs have an impact on society. Further, they highlighted themes that were of concern to young women such as work-life balance and attempted to demystify these through discussions and sharing experiences of women working in these domains. In addition to these videos, Ève's team

---

[19] "Chairholder Profile: Ève Langelier", May 20, 2014, accessed May 12, 2019, http://www.nserc-crsng.gc.ca/Chairholders-TitulairesDeChaire/Chairholder-Titulaire_eng.asp?pid=912; "Women in Physics—Ève Langelier", July 2018, accessed May 12, 2019, https://www.physique.usherbrooke.ca/wipc/en/eve-langelier-4/ and Ève Langelier et Marcela Cid, "La promotion des carrières en STIM auprès des filles", in Jeanne d'Arc Gaudet and Louise Lafortune (eds.), *Les grands enjeux des femmes pour un développement durable*, Sainte-Foy, Presses de l'Université du Québec, 2010, p. 161–173.

[20] Ève Langelier, *Rapport d'étape de 24 mois (mai 2015 à avril 2017)*, Sherbrooke, Université de Sherbrooke, January 2018, p. 8; Ève Langelier, Interview with Hannah Elizabeth Young, telephone interview, May 31, 2018, and "Formation sur la démarche de conception", date unknown, accessed May 12, 2019, http://cfsg.espaceweb.usherbrooke.ca/formation-demarche-conception/

translated into French eight informative white papers on gender diversity, initially produced by Westcoast Women in Engineering, Science and Technology (WWEST, Elizabeth Croft's Chair) in 2014. Alongside the creation of these resources, the Quebec Chair also raised public awareness through interviews given to the media resulting in many articles in newspapers and magazines, and radio episodes. Finally, Ève and her team gave more than 50 presentations about the underrepresentation of women in science and engineering at various conferences, congresses, workshops, and round tables in Quebec, Canada, and around the world.[21]

During her first term as Chair, Ève developed several new programs and leant support to projects already in existence. As previously mentioned, she had been involved with the *Les filles et les sciences, un duo électrisant*! event since 2007. Upon taking up the Chair position, she decided to take it a step further. Starting in 2016, she used her funding to financially support the event in Sherbrooke and helped with the recruitment of volunteers and developing workshops. In 2016, the Eastern Townships event attracted 50 girls of Secondary 2 and 3 while the 2017 edition brought together 70 girls of Secondary, 2, 3, and 4. This day-long event has been a success since then. According to a participant survey, 85% of the 2017 edition's participants in the *littleBits Kit* workshops, which focused on basic notions in electronics and design, enjoyed the activity and thought it was useful for their future training in science and technology.[22]

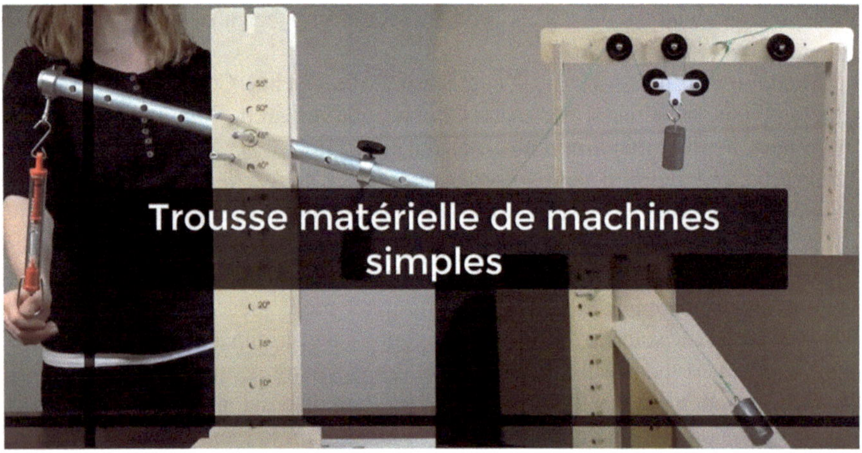

[21] Ève Langelier, *Rapport d'étape de 24 mois (mai 2015 à avril 2017)*, Sherbrooke, Université de Sherbrooke, January 2018, p. 1–4; "Série 'Changer au féminin'", date unknown, accessed May 12, 2019, http://cfsg.espaceweb.usherbrooke.ca/serie-changer-monde-feminin/; "Série 'D'une femme à une autre'", date unknown, accessed May 12, 2019, http://cfsg.espaceweb.usherbrooke.ca/serie-dune-femme-a-une-autre/; "Feuillets sur la diversité des genres", date unknown, accessed May 12, 2019, http://cfsg.espaceweb.usherbrooke.ca/feuillets-sur-la-diversite-des-genres/ and "Conférences et présentations", date unknown, accessed May 12, 2019, http://cfsg.espaceweb.usherbrooke.ca/conferences-et-presentations/

[22] Ève Langelier, *Rapport d'étape de 24 mois (mai 2015 à avril 2017)*, Sherbrooke, Université de Sherbrooke, January 2018, p. 3.

The Quebec Chair also created various resources to support the work of primary and high school teachers in science and technology, with the help of Prof. of Education Fatima Bousadra. First, the Chair's staff produced the interdisciplinary didactic kit *Safari-photo (hors de l'ordinaire) au Labo-Sphère* ("Safari-photo (out of the ordinary) in the Lab-Sphere) for primary school teachers. The kit aimed to build positive science and technology experiences for young girls in order to give them the opportunity to consider, without bias or lack of confidence, the possibility to study and work in these fields. Second, the Chair developed the *Trousse matérielle de machines simples 3 en 1* ("3 in 1 material kit of simple machines") to provide low-cost materials actually used by engineers to primary and high school teachers. The kit allowed students to experiment in class with levers, pulleys, and inclined planes by varying the different parameters. In addition to acting as a demonstrator, the kit could also be used to create an authentic environment. A series of explanatory video capsules was provided for free on the Chair's webpage. Tested by 190 undergraduate students in preschool and primary schools teaching and 70 Grade-6 students, the kit was one of the best outcomes of the Chair in its early years. Indeed, it allowed significant learning in technology, enhanced students' interest, and allowed a better understanding of the engineering profession. Furthermore, it was downloaded 156 times in only two years. Third, the *Formation sur la démarche de conception*, training materials and in-person training on the design process aimed at high school teachers, was developed by Ève's team as a means to help teachers improve their own teaching in "the technological world" and enable them to develop the cognitive pleasure of solving technological problems in girls (and boys). The Chair's team created self-led and in-person training modules which used a realistic scenario, such as the design of a launching device as a swimming pool toy, to demonstrate the different steps of the engineering design approach. An accompanying series of video capsules was provided to explain the steps of the design process. These videos presented theoretical aspects accompanied by real-life examples from various engineering fields. In 2017, Ève started teaching a course to future high school teachers: GET201: Introduction à la conception technologique (Introduction to Engineering Design).[23]

In addition to her work for primary and high school teachers, Ève and her team also worked on recruitment, retention, and promotion strategies aimed at women CEGEP and university students and professors. For women high school and CEGEP students, she created a webpage on the Academos cybermentoring platform and

[23] Chaire pour les femmes en sciences et en génie, *Démarche de conception technologique: Vers la construction d'un cahier des charges fonctionnel (CdCF)*, Sherbrooke, Chaire pour les femmes en sciences et génie, 2017, 25 p., http://cfsg.espaceweb.usherbrooke.ca/download/3172/; Ève Langelier, *Rapport d'étape de 24 mois (mai 2015 à avril 2017)*, Sherbrooke, Université de Sherbrooke, January 2018, p. 2; "Trousse matérielle de machines simples", date unknown, accessed May 13, 2019, http://cfsg.espaceweb.usherbrooke.ca/trousse-materielle-de-machines-simples/; "Trousse interdisciplinaire", date unknown, accessed May 13, 2019, http://cfsg.espaceweb.usherbrooke.ca/trousse-interdisciplinaire/ and "Démarche de conception", date unknown, accessed May 13, 2019, http://cfsg.espaceweb.usherbrooke.ca/demarche-de-conception/

supplied it with information on science and engineering studies and careers, to demystify science and engineering, promote careers in these fields, and offer connections with women role models through mentoring relationships. Through this social network, the Chair reached more than 90,000 young people from across the province and worked in collaboration with more than 510 registered schools. As of 2019, Ève's team recruited nearly 50 mentors through their Academos page. For women students and professionals in the science and engineering fields, she organized a leadership workshop and a communications workshop (*Devenir leader* and *Communiquer efficacement*), both translated and adapted from WinSETT workshops available in English across most of Canada (*Becoming Leaders* and *Effective Communication*, respectively). By offering these training workshops twice in 2017, Ève wanted to help these women continue their education and progress in their profession, as well as to give them the opportunity of networking with each other.[24]

Among the three major studies she undertook to understand the question of women in science and engineering in Quebec, Ève published reports based on the analysis of enrollment data for women in undergraduate, Master's, and doctoral programs in Quebec between 2005 and 2019. The statistical data has been updated each year of the Chair's mandate and published on her website. The 2019 publication, covering Quebec's enrollments from 2007 to 2018, showed that the participation of women in engineering enrollments increased at all three university levels: undergraduate (by 5%), Master's (by 7%), and doctorate (by 6%). Despite this improvement, women remained underrepresented in engineering fields. In science, the results showed that very little had changed over these 10 years: participation by women in undergraduate programs held steady (at 43%), while it decreased slightly in Master's programs (−1%) and increased slightly in doctorate programs (+ 2%). While women's enrollment reached 30% or more in science in general, women remained underrepresented in certain fields of study such as computer science, physics, and mathematics.[25]

[24] Ève Langelier, *Rapport d'étape de 24 mois (mai 2015 à avril 2017)*, Sherbrooke, Université de Sherbrooke, January 2018, p. 3; "Cybermentorat Academos", date unknown, accessed May 13, 2019, http://cfsg.espaceweb.usherbrooke.ca/cybermentorat-academos/ and "Ateliers de formation", date unknown, accessed May 13, 2019, http://cfsg.espaceweb.usherbrooke.ca/ateliers-de-formation/

[25] Ève Langelier, *Rapport d'étape de 24 mois (mai 2015 à avril 2017)*, Sherbrooke, Université de Sherbrooke, January 2018, p. 3; "Statistiques", date unknown, accessed May 13, 2019, http://cfsg.espaceweb.usherbrooke.ca/statistiques/

The Ontario Chair: Catherine Mavriplis (2011–2021)

When Catherine Mavriplis began her mandate as the Pratt & Whitney Canada/NSERC Chair for Women in Science and Engineering for Ontario in 2011, she had been a professor in the Department of Mechanical Engineering at the University of Ottawa for only three years. Yet, she had extensive experience as a faculty member and researcher. She had worked as a professor of mechanical and aerospace engineering for 20 years, mostly in the United States, including a two-year rotation as a program manager at the US National Science Foundation (NSF). Catherine was involved in numerous research projects related to advanced numerical methods for fluids dynamics and applications in aerodynamics and other interdisciplinary fields such as combustion, meteorology, and blood flow. Although she always relished studying numerical methods for high-fidelity aerodynamics simulations, Catherine had also been a tireless advocate for promoting women in STEM for a long time. Having experienced the wage gap through an employment offer from the Boeing aerospace company (that she turned down) and later the lack of maternity leave in academia in the United States, she understood that women in these fields needed support to overcome these obstacles. Since 1996, she has worked under NSF funding to advance women's issues, notably through the *FORWARD program* she developed with colleagues and the NSF ADVANCE programs, reaching over 2,500 women by 2011. For 15 years, she had been doing the same kind of work as a Chair, but in the United States. Her first encounter with the NSERC Chairs program happened at the 2002 ICWES12 conference organized by the first regional Chairs Monique Frize and Claire Deschênes. From then on, she knew that she wanted to come back to Canada and apply for the position.[26]

[26] "Catherine Mavriplis", date unknown, accessed June 29, 2019, https://gender-summit.com/gs11-na-2017/gs11-info-in-english/who-are-the-panelists/1745-catherine-mavriplis; "Catherine Mavriplis", date unknown, accessed June 29, 2019, https://engineering.uottawa.ca/people/mavriplis-catherine; Janice Koch, Berverly Irby, and Barbara Polnick (eds.), *Girls and Women in STEM: A Never Ending Story*, Charlotte, N.C., Information Age Publishing Inc., 2014, p. 245; Catherine Mavriplis, Interview with Hannah Elizabeth Young, September 8, 2018 and Catherine Mavriplis's testimony to the House of Commons Standing Committee on the Status of Women, 1st Session, 42nd Parliament, May 9, 2017, accessed July 15, 2019, https://www.ourcommons.ca/DocumentViewer/en/42-1/FEWO/meeting-60/evidence

Catherine designed her Ontario Chair program around three observations she had made. First, it seemed to her that there were many programs targeting kids and university students. For example, by 2011, Ontario Network of Women in Engineering (ONWiE) was well on its way to standardizing outreach programs across the province. So, Catherine understood from the start that she should concentrate her attention on a different area. Second, she had a keen interest in leadership development and thought she could focus her attention on mid-career development and access to leadership roles. In fact, the Ontario Chair quickly realized that, even though more women entered the pipeline, the yield at higher levels was stalled. Surely, there was something preventing them from reaching these positions. As a result, Catherine wanted to understand how women persisted and how they were held back, and this oriented part of her program.[27] Finally, the 25 years Catherine spent in the United States had a profound impact on her. Like the former West Coast Chair Anne Condon, Catherine appreciated NSF's approach to increasing the participation and advancement of female scientists and engineers through a closer look at institutions and their processes. Their requirement to include interdisciplinary research had a special appeal to her. During her time at George Washington University, she was asked to attend several university-level committees. She quickly learned that she enjoyed working with people from other department and faculties because colleagues who choose to be involved with other disciplines usually have a more open mind, she observed. It seemed paramount to include this aspect into her own research on women in science and engineering.[28]

[27] Catherine Mavriplis, Interview with Hannah Elizabeth Young, August 9, 2018.

[28] Catherine Mavriplis, Interview with Hannah Elizabeth Young, August 9, 2018.

The Ontario Chair program addressed three specific areas: recruitment to the professions, retention and advancement of professionals already employed in STEM, and interdisciplinary research on the status of women and the science and engineering environment. Recruitment to university programs was well served by existing outreach programs at most universities: they seemed to be successful at introducing science and engineering to young girls. However, some gaps remained and Catherine sought to fill them with two key projects. First, Catherine observed that girls were not getting introduced to computer science or computer programming at a young enough age. Many students arrived at university with little or no programming experience, much like 30 years ago. The difference though was that our world had become increasingly driven by technology and women were missing out, since they could feel intimidated by those who did program for fun or in high school computer science classes. *Go Code Girl/Allez coder les filles* was inspired by *Go ENG Girl/GÉNiales les filles*, the outreach program initiated by the former Ontario Chair, Valerie Davidson, in 2005, to inspire girls to pursue engineering careers. Created in 2013, *Go Code Girl/Allez coder les filles* aimed to teach computer programming to girls, introduce them to the world of software development, and discover opportunities in computing and engineering fields. She worked with the Faculty of Engineering at the University of Ottawa and the School of Computer Science at Carleton University to develop a curriculum that was relevant and appealing to girls in Grades 7–11. The first two-day interactive workshop initiated 22 participants to basic programming and Web development and offered them an opportunity to play with hardware such as Raspberry Pis and Arduinos. Since 2015, the event has been supported by ONWiE and brought to other Canadian universities. By 2018, 12 Canadian universities were running the event, reaching a total of 630 girls across Canada. The program is now offered by 18 universities across Canada, serving 800 girls a year. Viewed as a tremendous success, the program was mentioned by the Honorable Kirsty Duncan, the federal Minister of Science, at the 2017 NSERC Awards for Science Promotion ceremony.[29]

Second, Catherine saw a gap in reaching Indigenous students, to attract them to science and engineering programs. She partnered with the Verna J. Kirkness Science and Engineering Education Program to heighten awareness among First Nations, Métis, and Inuit youth about the various careers and academic paths related to STEM fields and increase the number of Indigenous students graduating from science and engineering programs. The program was inspired by the work of the

[29] Catherine Mavriplis, Interview with Hannah Elizabeth Young, August 9, 2018; ONWiE, "Go Code Girl", date unknown, accessed July 24, 2019, http://www.onwie.ca/programs/go-code-girl/; Nathalie Vallières, "Celebrating girls who code", *University of Ottawa Gazette*, March 23, 2015, accessed July 24, 2019, https://www.uottawa.ca/gazette/en/news/celebrating-girls-who-code; CWSE-ON, "Outreach", date unknown, accessed July 24, 2019, http://cwse-on.ca/programs/outreach; NSERC, "Canada's science champions receive top honours from the Government of Canada", May 15, 2017, accessed July 24, 2019, http://www.crsng-nserc.gc.ca/Media-Media/ProgramNewsDetails-NouvellesDesProgrammesDetails_eng.asp?ID=912 and ONWiE, *2018 Report*, Guelph, ONWiE, 2018, p. 3, accessed July 24, 2019, http://www.onwie.ca/wp-content/uploads/2019/02/2018-Report-2-compressed.pdf

Prairie Chair, Annemieke Farenhorst, who was already involved with the Kirkness Foundation. Beginning in 2009, the program offered scholarships to First Nations, Métis, and Inuit Grade-11 high school students to spend a week at a Canadian university to explore science or engineering as a member of a research team. During this time, students had the opportunity to conduct experiments and attended lab meetings, learned about the support systems that were available to them on campus, and met role models. Seeing the great benefits of this initiative, Catherine decided to support the program at her home institution. The University of Ottawa's program would be the first expansion of the Kirkness program to the East. Her call to fellow professors to welcome trainees at the University of Ottawa was enthusiastically answered by 20 professors. The one-week research visit saw the participation of 25 girls and three boys over 2018 and 2019 who worked with 13 professors in science and engineering and their graduate students. The experience allowed Catherine to build important ties to the community through the Kirkness Foundation, the University of Ottawa Indigenous Resource Centre, the Wabano Centre for Aboriginal Health in Vanier, and Indigenous elders in the community. Canadian Broadcasting Corporation (CBC) Radio came on campus to interview the "Kirkness scholars" as they were named, and the story was well received by the Ottawa-Gatineau community as reported back to the Chair informally. The interview provided an awareness and exposure of Indigenous girls' capabilities and future goals for themselves and their communities.[30]

[30] Catherine Mavriplis, *24-month Report*, Ottawa, University of Ottawa, June 2019, p. 1–2; CWSE-ON, "Outreach", date unknown, accessed July 24, 2019, http://cwse-on.ca/programs/outreach; Kirkness Foundation, "Verna J. Kirkness Science and Engineering Education Program", 2019, accessed July 14, 2019, http://www.vernajkirkness.org/program-description/ and Radio-Canada Première, "De jeunes autochtones expérimentent la recherché scientifique", May 17, 2018, accessed July 24, 2019, https://ici.radio-canada.ca/premiere/emissions/Les-matins-d-ici/segments/entrevue/72399/science-autochtones-universite-ottawa

In terms of retention and advancement of professionals, the Ontario Chair created three different workshops to develop professional and leadership skills for working women in science and engineering. First, a *Negotiating in a Male-Dominated Environment* workshop was developed with Communications Master's student Jerie Shaw and offered 15 times in nine different cities reaching a total of 500 people, including within the industrial sponsor site, at Pratt & Whitney Canada. An analysis of the effectiveness of the negotiation workshops showed 88% of those surveyed felt more confident to negotiate and 100% of those interviewed reported being able to apply the skills learned at the workshop in the workplace. Additionally, two workshops for women professors were held. *Take the Final Step/Passez à l'étape finale* was developed and offered in 2013 to 45 women associate professors working toward promotion to full professorship. The model was then adopted and replicated by the Atlantic Chair and the British Columbia and Yukon Chair programs. A national *FORWARD to Professorship Canada* workshop for aspiring and current pre-tenure assistant professors, based on Catherine's previous work in the United States, was offered in 2015 to 40 participants. The National Network of NSERC Chairs for Women then adopted these two workshops to be delivered across the country. To date, over 3,600 women have participated in the United States and Canada.[31]

Catherine proposed in her Chair application and later developed several interdisciplinary research projects, teaming with researchers in communications, history, sociology, education, and management. Among the eight interdisciplinary projects launched by Catherine's team, three stand out in particular. The first one, entitled *De mère en fille: les femmes en STIM*, studied the influence of mothers on their daughters' choices to study STEM fields. A total of 14 professional women and 12 students were interviewed for this project. The work, co-supervised by Catherine and education professor Donatille Mujawamariya, showed that none of the participants questioned the role her mother played in her choice or in pursuing her studies or careers in these domains. For some, their mother was a model to reproduce, especially if she had studied or had a career in STEM. For others, their mother encouraged them to rush into these male-dominated fields. The collaboration with Education Prof. Mujawamariya later led to a new SSHRC-funded project in 2017 on women in engineering: le génie au service des femmes ("engineering in the service of women")—rethinking the faces and spaces of engineering.

The second research project explored gender differences in the early career experiences of engineering graduates in Canada by examining their entry into the occupation, job satisfaction, and their probability of looking for another job. Using quantitative methods to identify how gender shaped graduates' experiences, this

[31] Catherine Mavriplis, *24-month Progress Report*, Ottawa, University of Ottawa, September 2013, p. 2–4; Catherine Mavriplis, *48-month Report*, Ottawa, University of Ottawa, August 2015, p. 2 and Catherine Mavriplis and Françoise Moreau-Johnson, "Initiatives to Empower Women Faculty Members at University of Ottawa: Mothers See the Value in Participation", *Journal of the Motherhood Initiative*, vol. 6, no. 2 (2015), p. 183–184.

doctoral thesis study of Victoria Osten focused its attention on women who were successful in their career three years after graduating with a Bachelor's degree in engineering. The analysis, co-supervised by the Ontario Chair and sociology professors the late Ann Denis and Phyllis Rippey, found that women were finding jobs sooner than men, they were more satisfied with their jobs than men, and they were unlikely to want to leave like men. However, parenthood was still impacting women more than men. Visible minority status also increased the chances for graduates to look for another job.[32]

A final interdisciplinary research project, entitled *Breaking the Code: Canadian Women in Computer Science Education from 1945–1995*, was started as doctoral thesis research by Jennifer Thivierge under the co-supervision of history professor Ruby Heap and Catherine. This work considered the history of Canadian women in computer science from the Second World War (1939–1945) until the dot-com bubble in 1995. Through analysis of historical documents, and interviews from both men and women who participated in computer science programs, the project analyzed the creation of computer science programs at four Canadian universities. An interesting part of this work showed that keypunch girls, women who punched holes by hand or with keypunch machines in data cards that were fed into adding machines or tabulators that read the information stored on them, were at the bottom of the computer "food chain," an acceptable occupation for women at this time. This situation changed drastically in the 1970s, after men, seeing the job opportunities available to them, enrolled in Canada's university computer science programs and overtook the field. The women were left behind and were not retrained to program computers when the keypunch machines became obsolete. This is but one example of findings from the project explaining the low participation of women in computer science in Canada in the cited years, a situation which persists to this day.[33]

[32] Victoria Osten, *Gender Differences in the Early Career Experiences of Engineers in Canada*, Ph.D. Thesis, University of Ottawa, 2019, 106 p.; Catherine Mavriplis, *48-month Report*, Ottawa, University of Ottawa, August 2015, p. 4–5 et Donatille Mujawamariya, Catherine Mavriplis, and Michelle Boucher, "De mère en fille: Les femmes en STIM (Sciences, Technologie, Ingénierie et Mathématiques), *Bulletin de l'AFFESTIM*, vol. 11, no. 1 (May 2015), p. 9, accessed July 24, 2019, http://www.affestim.org/fileadmin/affestim/intranet-affestim/Documents_Officiels/Autres/InfoAFFESTIM2015.pdf

[33] Catherine Mavriplis, *48-month Report*, Ottawa, University of Ottawa, August 2015, p. 4–5; CWSE-ON, "Current Projects", date unknown, accessed July 24, 2019, http://www.cwse-on.ca/research-1/current-projects and Linda Scales, "Researching the untold story of Canada's keypunch girls", *University of Ottawa Gazette*, October 9, 2018, accessed July 24, 2019, https://www.uottawa.ca/gazette/en/news/researching-untold-story-canadas-keypunch-girls

The Prairie Chair: Annemieke Farenhorst (2011–2020)

When Annemieke Farenhorst began her mandate as the Chair for Women in Science and Engineering for the Prairie Region in 2011, she had been a professor in the Department of Soil Science at the University of Manitoba for 14 years. Annemieke's research interests lay in pesticides, natural steroid estrogens, and antibiotics in soil and water as well as drinking water quality in First Nations communities. As the first woman professor appointed in her department, she always felt isolated, not because of her male colleagues. Annemieke always considered she was in a very good department with great colleagues, but being able to connect with at least one other woman in science and engineering would have lessened this feeling of social isolation. Her interest in the Chair was thus sparked partly from her own professional condition in her department. But it also came from the Associate Dean of Research in the Faculty of Agricultural and Food Sciences at the University of Manitoba when she asked Annemieke if she would be interested in this position. Knowing nothing about the Chair program, she looked at it and, from the start, Annemieke knew that the work would be really captivating.[34]

Annemieke's program partly bore the mark of her own professional identity. As a faculty member in the Faculty of Agricultural and Food Sciences, she was one of the only women professors teaching at the undergraduate level. Yet, the number of women undergraduate students in the faculty was slightly greater than 50%. Thus, her own faculty setting prompted her interest in the presence of role models for women students as well as the workplace environment of women academics in science and engineering. Annemieke was also influenced by the Prairie provinces' demographics. Like the former Prairie Chair, Julita Vassileva, Annemieke understood quickly that she needed to target the participation of Indigenous girls and women. In the provinces of Saskatchewan and Manitoba alone, about 20% of the preschool population was Indigenous. To be really inclusive in her program, she needed to create opportunities specifically for Indigenous youth. The focus on First Nations communities was further reinforced by the funding from the Verna

[34]Annemieke Farenhorst, Interview with Hannah Elizabeth Young, telephone interview, June 6, 2018.

Fig. 4.2 Prairie Chairholder Annemieke Farenhorst collecting water samples from Island Lake, Manitoba, with Waylon Hill and Elder Russel Harper as part of the NSERC CREATE H2O project she led

J. Kirkness Education Foundation.[35] One of the major sponsors of the Chair, this organization was actively engaged in the Chair's vision "to increase the number of Indigenous students graduating from science and engineering programs in Canada."[36]

Annemieke's program focused on three key areas: recruitment in STEM professions, retention and networking, and the academic workplace environment and support to women academics. First, among various recruitment activities, the Prairie Chair put a special emphasis on supporting Indigenous girls in the region. In that regard, she started the *CREATE H2O program for First Nations Water and Sanitation Security* in 2013. The project was funded by a $1.65-million NSERC CREATE grant over six years with matching funds from Trent University, University College of the North, the Assembly of First Nations, and industry partners (Fig. 4.2).

The graduate training project addressed the long-standing lack of accessible sanitary drinking water within First Nations communities. By focusing on building research collaborations with First Nations communities to evaluate wastewater treatment systems, as well as spring water and drinking water quality, this interdisciplinary research and training initiative enabled students to add to their academic achievements and job skills that are valuable in the labour market. Annemieke's

[35] Annemieke Farenhorst, *2019 Progress Report*, Winnipeg, University of Manitoba, 2019, p. i and Annemieke Farenhorst, Interview with Hannah Elizabeth Young, telephone interview, June 6, 2018.

[36] NSERC, "Chairholder Profile: Annemieke Farenhorst", date unknown, accessed July 25, 2019, http://www.nserc-crsng.gc.ca/Chairholders-TitulairesDeChaire/Chairholder-Titulaire_eng.asp?pid=770

team was responsible for overseeing most aspects of the training program, including program curriculum; programming related to the recruitment and advancement of students in the program; building and maintaining partnerships with stakeholders; and financial management and accountability. Between 2013 and 2019, the program trained 100 students and postdoctoral fellows (of whom 33% were Indigenous and 63% were women), and they collectively worked with 30 First Nations communities in Manitoba, Ontario, and Saskatchewan.[37]

As we know, the Prairie Chair has been involved with the Verna J. Kirkness Science and Engineering Education Program since 2012. The partnership with the Kirkness Foundation aimed to heighten awareness among First Nations, Métis, and Inuit youth about the various careers and academic paths related to STEM fields and increase the number of Indigenous students graduating from science and engineering programs. Annemieke assisted the Kirkness Foundation by leading survey design and data collection, designing annual impact reports, recruiting youth and mentors to the program (including youth in remote communities), developing a communication strategy to share program information more broadly (including with youth in communities with no or little wi-fi access), providing on-site human resource assistance to the program operations at four universities (facilitating evening and extracurricular activities for the students, supervising students in their dorms, and visiting and assisting with lab activities), sponsoring travel of all girls visiting universities in the Prairie provinces, and girls from the Prairies visiting other universities, and, finally, mentoring girls visiting the Chair's research team and laboratories. If the program was first only offered at the University of Manitoba, it was quickly joined by the University of Saskatchewan in 2015 and expanded to eight university partners in 2019. In 2019 alone, the program involved 141 youth (of whom 62% were girls) in Grade 11 and their chaperones. Of these, 54 were sponsored by the Chair, and four girls and one boy worked for one week in Annemieke's laboratories where they learned to do soil and water sampling, chemical and microbial analyses, and studied soil organisms. Since its inception in 2009 to 2018, the Kirkness program has supported 512 Indigenous students from more than 109 communities across Canada. The surveys conducted showed that 80% of the youth graduating from this program entered universities afterward, with the majority selecting science and engineering fields. Annemieke's strong commitment to First Nations development was also reflected in her own lab, where 16% of her recent research trainees self-identified as Indigenous and were enrolled in undergraduate and graduate programs.[38]

---

[37] "NSERC CREATE H20 Program for Frist Nations Water and Sanitation Security", date unknown, accessed July 26, 2019, http://create-h2o.ca/, Alycia Rodrigues, "CREATE H20: First Nations solutions to First Nations Issues", *The Manitoban*, September 2, 2013, accessed July 26, 2019, http://www.themanitoban.com/2013/09/create-h2o/16071/; Press release, "Trent University to Aid First Nations Water Management in National $1.6 Million Program", *Indigenous Lands & Resources Today*, August 28, 2013, accessed July 26, 2019, http://www.ilrtoday.ca/trent-university-to-aid-first-nations-water-management-in-national-1-6-million-program/ and Annemieke Farenhorst, *2019 Progress Report*, Winnipeg, University of Manitoba, 2019, p. i and Appendix B.

[38] Annemieke Farenhorst, *2019 Progress Report*, Winnipeg, University of Manitoba, 2019, p. i, 2–3 and Appendix C.

Addressing retention, two major activities stand out. The first one was the production of videos featuring female role models from a wide range of agricultural professions. The purposes of the videos were to show future students, first, that they could become or be a parent and earn a graduate degree at the same time and, second, that people can excel as students despite challenges. The videos were made available online and were also shared at a range of events. These resources were also integrated as permanent displays in the first hands-on facility to explore the ways in which food is made in Canada, the Bruce D. Campbell Farm and Food Discovery Centre (FFDC), located at the National Centre for Livestock and the Environment, at the University of Manitoba's Glenlea Research Station in Winnipeg. The official launch was attended by more than 30 stakeholders. The FFDC attracts more than 5,000 visitors each year, of which 50% were teachers and youth. Typically, the feedback was very positive, and numerous people said the videos were inspiring.[39]

In 2017, Annemieke decided to establish the *Faculty of Agricultural & Food Sciences (FAFS) Mentorship Program*, modeled after the success of its sister program, the Committee for Increasing the Participation of Women in Engineering (CIPWIE) Engineering Mentorship Program, an operating committee of Engineers Geoscientists Manitoba (the professional association to regulate the profession of engineering in Manitoba). The latter program was developed by the CIPWIE to respond to the 30 by 30 goal, Engineers Canada's commitment to raise the percentage of newly licensed engineers who are women to 30% by the year 2030. Since 2015, the CIPWIE Engineering Mentorship Program has paired women engineering students with women professional engineers. The goal was to aid in the retention of female engineering talent in the long term by providing students with an opportunity to see what a career in engineering could look like. The program was partly possible through the sponsorship of Annemieke's Chair. Building on this success, the Prairie Chair started the *FAFS Mentorship Program* to match up university students with industry and academic mentors in their field. Over the course of this eight-month program, the students and the mentors built relationships through a combination of regular smaller-group meetings and several more structured events that brought together all program participants. In these organized group events, the participants were encouraged to discuss a common topic or specific learning area. Through this initiative, students had the opportunity to explore career goals, network with women in the field, and develop new skills under the guidance of their mentors. In the first two years of its existence, the *FAFS Mentorship Program* brought together 88 mentors and 42 students. Again, the feedback both from mentees and mentors was overwhelmingly positive.[40]

---

[39] Annemieke Farenhorst, *2019 Progress Report*, Winnipeg, University of Manitoba, 2019, p. i and 4.

[40] "2016–2017 Committee for Increasing the Participation of Women in Engineering (CIPWIE)", 2017, accessed July 26, 2019, http://www.apegm.mb.ca/Ingenium/2017/2017/ReportCIPWIE.pdf; Kathryn Atamanchuk, Kristen Myles, and Morgann Becket, "Mentorship: Linking Industry Professionals with Tomorrow's Engineers" in *Proceedings of the 2016 Canadian Engineering Education Association (CEEA16) Conference*, Paper 057, Dalhousie University, Halifax, Nova

Turning to the academic workplace environment and support to women academics, Annemieke developed and launched a workplace climate survey in collaboration with four social scientists. The survey was administered to 421 Natural, Sciences and Engineering faculty from five universities in the Prairie Region. As in the survey conducted in the Atlantic, the results for the Prairies showed that faculty members in these fields were generally optimistic about the climate in their own departments or units when compared to the overall university climate and about their own career progress as well. The data showed, however, that women faculty members experienced more negative day-to-day workplace experiences like harassment and discrimination and had higher levels of workplace stress than their male counterparts. The summary report was distributed to all respondents, administrators at the participating universities, as well as with the government of Canada's Office of the Minister of Science and Sport, and four manuscripts have been submitted to academic publications.[41]

The West Coast Chair: Elizabeth Croft (2010–2015)

Scotia, Canada, June 19–22, 2016, unknown date, accessed July 26, 2019, https://ojs.library.queensu.ca/index.php/PCEEA/article/view/6444/5992; "Agriculture Mentorship Program", date unknown, accessed July 26, 2019, http://cwse-prairies.ca/pages/mentorship/mentorship.php and Annemieke Farenhorst, *2019 Progress Report*, Winnipeg, University of Manitoba, 2019, p. i and 3–4.

[41] Annemieke Farenhorst, *2019 Progress Report*, Winnipeg, University of Manitoba, 2019, p. i and 5 and Annemieke Farenhorst, Tracey Peter, Jennifer Dengate, Danielle Saj, Sally Marchand, and Mahalia Lepage, *Faculty Workplace Climate Survey Results—Prairie Region Report*, Winnipeg, NSERC Chairs for Women in Science and Engineering, 2018, p. 3–5.

At the time of her appointment as NSERC Chair for Women in Science and Engineering for British Columbia and the Yukon (Westcoast Women in Engineering, Science and Technology (WWEST)) in 2010, Elizabeth Croft had been a professor in the Department of Mechanical Engineering at the University of British Columbia (UBC) for 15 years. She had also been serving as the Associate Head of her department for three years where she was in charge of student recruitment and external affairs. As director of the Collaborative Advanced Robotics and Intelligent Systems (CARIS) Laboratory at UBC, a world-class robotics lab, Elizabeth has made significant contributions to the field of robotic systems. Through collaborative research projects with General Motors, the Deutsches Zentrum für Luft- und Raumfahrt (DLR - the German Aerospace Centre), and other industry partners, she studied the interaction of robots and humans for manufacturing assembly, health care, and assistive technology. If she did encounter some difficulties as a woman student and professional in engineering herself, Elizabeth's interest in women's issues in science and engineering was sparked mostly by external influences. As one of the few women professors in her department, she was approached by a group of women students in 2003 to start a mentoring program. The success of the pilot program led male students to ask for the same opportunity. Eventually, she was able to get funding from the faculty to support a program for everybody. Then, in 2007, she received funding from the former West Coast Chair program, led by Anne Condon, to start "Building Communities," which is now known as the *Creating Connections* conference series. When Anne decided that she was not going to renew as Chair, she strongly encouraged Elizabeth to apply for it. Other colleagues also convinced her that she should apply stating that, with this type of funding, she could make a real and sustainable impact.[42]

Elizabeth's work as the West Coast Chair was influenced by three main factors. First, the fact that the Chair was sponsored by multiple donors was both an inconvenience and an advantage. It was an inconvenience, because she had to respond to all their varied concerns. Yet, it was also an advantage, since she was able to really set her own agenda unlike other Chairs who were sponsored by a single organization. Indeed, none of her sponsors had a large enough impact to tell her what to do. Second, her research collaboration with several industry partners made her well acquainted with the problems encountered in this sector. She learned that many women were actually leaving the engineering profession, burned out by the workplace climate and permeating biases at play in a male-dominated engineering

[42] "Elizabeth Croft", date unknown, accessed May 15, 2019, https://research.monash.edu/en/persons/elizabeth-croft; "Professor of Mechanical Engineering and Associate Dean, Education and Professional Development for the Faculty of Applied Science at UBC", date unknown, accessed May 15, 2019, https://www.scienceworld.ca/elizabeth-croft; Anne Millar and Mary Wells, "Elizabeth Croft Brings Women in STEM Together", *CIM Magazine*, September 14, 2017, accessed May 15, 2019, http://magazine.cim.org/en/voices/elizabeth-croft/ and Elizabeth Croft, Interview with Hannah Elizabeth Young, telephone interview, September 10, 2018.

culture. The persistent gender wage gap between men and women was also another problem she was very much aware of since she had studied this specific issue prior to her appointment as Chair. Finally, the presence in British Columbia and Yukon of various organizations dedicated to the promotion of women in STEM meant that Elizabeth could have a much bigger impact as Chair if she partnered with them instead of trying to do it all by herself. This was the reasoning behind her willingness to build a community of practice and provide financial and expertise resources to them.[43]

In addition to her work as Associate Dean in the Faculty of Applied Science from 2013 onward, the West Coast Chair focused her time on awareness and outreach, recruitment and retention, as well as research. Among her awareness and outreach achievements, the creation of the Westcoast Women in Engineering, Science and Technology (WWEST) Partners Program in 2011 emerged as one of her most striking initiatives. Inspired by Anne Condon's Jade Bridges Programme, Elizabeth felt that she should provide funding to new and established not-for-profit organizations advancing gender diversity in science and engineering instead of trying to launch yet another project. However, she believed that the network should not just bring these partners together and give them financial resources. To deliver meaningful results, the WWEST Partners Program should also support their success and long-term sustainability through targeted training opportunities and professional development. It also asked them to collaborate and share best practices and participate in face-to-face networking opportunities. In the course of five years, Elizabeth funded twenty not-for-profit and university-based organizations throughout British Columbia and Yukon to create new programs or expand existing ones, build a sustainable plan, and collaborate more and more effectively throughout the network. The WWEST Partners Program was such a success that Lesley Shannon, the West Coast Chair from 2015 to 2020, would continue it.[44]

Among the recruitment and retention activities developed by the West Coast Chair, the first step was to focus on "publicly promoting the range of opportunities

[43] Elizabeth Croft, Interview with Hannah Elizabeth Young, telephone interview, September 10, 2018; Anne Millar and Mary Wells, "Elizabeth Croft Brings Women in STEM Together", *CIM Magazine*, September 14, 2017, accessed May 15, 2019, http://magazine.cim.org/en/voices/elizabeth-croft/ and Elizabeth Croft, *Final Report: NSERC Chair for Women in Science and Engineering BC and Yukon Region (2010–2015)*, Vancouver, University of British Columbia, 2015, p. 1–3 and Appendix 12, accessed May 15, 2019, http://wwest.sites.olt.ubc.ca/files/2015/09/WWEST-Final-Report-with-Appendices.pdf

[44] Anne Millar and Mary Wells, "Elizabeth Croft Brings Women in STEM Together", *CIM Magazine*, September 14, 2017, accessed May 15, 2019, http://magazine.cim.org/en/voices/elizabeth-croft/ and Elizabeth Croft, *Final Report: NSERC Chair for Women in Science and Engineering BC and Yukon Region (2010–2015)*, Vancouver, University of British Columbia, 2015, p. 3 and Appendix 12, accessed May 15, 2019, http://wwest.sites.olt.ubc.ca/files/2015/09/WWEST-Final-Report-with-Appendices.pdf

in STEM careers, the collaborative nature of the professions, and the ability for STEM professionals to create positive change in local and global communities."[45] The objective was to align with the overall NSERC CWSE targets to attract women to STEM careers, increase the enrollment of women in STEM programs at universities, and eliminate existing barriers. To meet this challenge, Elizabeth and her team changed the way they talked about STEM. They presented engineering as a creative, engaging, and rewarding profession. They put forward a profession where people solve problems, design solutions, and help both local and global communities. By changing the way they talked about STEM, the Chair's team was able to increase the proportion of women in the first-year engineering class at UBC from 20 to 30 percent between 2011 and 2015.[46]

Second, to ensure that outreach efforts to bring more young women into engineering would continue beyond Elizabeth's tenure as Chair, and to address a shortage of approximately 100,000 engineers who had been predicted by Engineers Canada by 2020, the Faculty of Applied Science and Goldcorp partnered to establish the Goldcorp Professorship for Women in Engineering at UBC in 2014. This initiative grew out of Goldcorp's earlier efforts to help women across the company unlock their potential, build skills, as well as achieve personal and professional fulfillment. This former program helped over 1,000 women to improve their self-esteem, increase their communication skills, and set new goals for themselves. With a $500,000 gift from Goldcorp, the objective of the Goldcorp Professorship was to increase the representation of women in undergraduate engineering programs from the national average of 20% in 2014 to 50% five years later. To accomplish that, the focus would be on promoting engineering as a creative and rewarding career to high school students, parents, and counselors. By promoting a profession where someone can make a real difference, the Goldcorp Professorship in Women in Engineering at UBC was thus the next step for the gold mining company to support diversity and bring more women into the profession across the nation. Dr. Sheryl Staub-French, associate professor in civil engineering, was selected as the inaugural holder of the Goldcorp Professorship. During her tenure, Sheryl created *Eng•cite*, a program promoting engineering education for teachers, counselors, parents, and high school students. Partnering with Geering Up, UBC's engineering outreach program, *Eng•cite*, provided engineering and science workshops to nearly 25,000 youth

[45] Elizabeth Croft, *Second 24-Month Progress Report (Months 24–48)*, Vancouver, University of British Columbia, 2014, p. 8.

[46] Elizabeth Croft, *Second 24-Month Progress Report (Months 24–48)*, Vancouver, University of British Columbia, 2014, p. 8; Elizabeth Croft, *Final Report: NSERC Chair for Women in Science and Engineering BC and Yukon Region (2010–2015)*, Vancouver, University of British Columbia, 2015, p. 1 and Anne Millar and Mary Wells, "Elizabeth Croft Brings Women in STEM Together", *CIM Magazine*, September 14, 2017, accessed May 15, 2019, http://magazine.cim.org/en/voices/elizabeth-croft/

across the province in 2018, including 12,000 girls, 5,500 inner-city youth, and 2,700 Indigenous youth.[47]

A third recruitment and retention activity grew out of Elizabeth's interactions with business leaders and professionals, which showed that concise fact sheets on factors that impact women in STEM supported by full citations would strengthen her efforts. Thus, Elizabeth and her team started the *Gender Diversity 101* project through which they created 14 short informational papers on varied topics, among which we find unconscious bias, mentoring works, stereotype threat, understanding workplace diversity, gendered words in job advertisements, and microagressions. To provide aligned organizations the opportunity to share the infographic content, a co-branding agreement was given to them in perpetuity. As of September 2015, over 2,000 copies of the fact sheets had been distributed. Ève Langelier's team also participated by translating the materials into French and made them available through several Chair websites. Finally, the *Gender Diversity 101* publications were packaged in a self-published book and distributed to libraries, postsecondary institutions, and industry groups.[48]

[47] "Dr. Sheryl Staub-French name inaugural holder of Goldcorp Professorship in Women in Engineering at UBC", June 25, 2014, accessed June 10, 2019, https://www.civil.ubc.ca/news-events/2014/06/dr-sheryl-staub-french-named-inaugural-holder-goldcorp-professorship-women-engineering-ubc; "The Changing Face of Engineering", March 10, 2014, accessed June 17, 2019, https://www.3blmedia.com/News/Changing-Face-Engineering; Elizabeth Croft, *Final Report: NSERC Chair for Women in Science and Engineering BC and Yukon Region (2010–2015)*, Vancouver, University of British Columbia, 2015, p. 6 and "Sheryl Staub-French Honoured At YWCA Women Of Distinction Awards Gala," May 14, 2019, accessed June 17, 2019, https://engineering.ubc.ca/news/2019/05/sheryl-staub-french-honoured-at-ywca-women-of-distinction-awards-gala

[48] Elizabeth Croft, *Final Report: NSERC Chair for Women in Science and Engineering BC and Yukon Region (2010–2015)*, Vancouver, University of British Columbia, 2015, p. 10; "Gender Diversity 101", November 2, 2016, accessed June 24, 2019, http://wwest.mech.ubc.ca/diversity/; "Feuillets sur la diversité des genres", Date Unknown, accessed June 24, 2019, http://cfsg.espace-web.usherbrooke.ca/feuillets-sur-la-diversite-des-genres/ and Rebekah Parker, Jennifer Pelletier, and Elizabeth Croft, *WWEST's Gender Diversity in STEM: A Briefing on Women in Science and Engineering*, Vancouver, The University of British Columbia, 2015, 59 p.

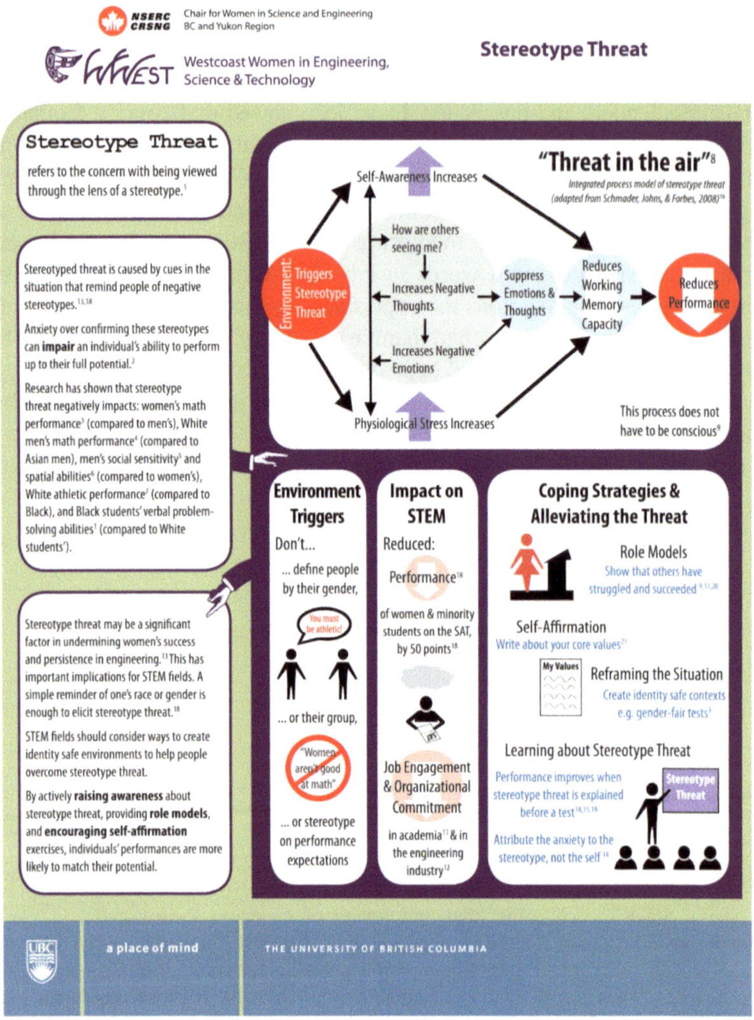

Regarding Elizabeth's research activities, the *Engendering Engineering Success* (EES) clearly stood out as an innovative and impactful project. Having worked closely with industry partners in her research around human-robot interaction, she saw countless women leaving the profession because of outdated stereotypes and negative workplace cultures that were still at play in Canadian organizations employing engineers. To understand and change aspects of workplace culture negatively affecting women, Elizabeth started collaborative research with Professor Toni Schmader, Canada Research Chair in Social Psychology, on the impact of workplace social interaction on the health and career satisfaction of both male and female professional engineers. She put together an interdisciplinary team of collaborators that included three Canada Research Chairs and a team of leading scientists from social psychology, business, and engineering. The group also collaborated with representatives from industry and industry associations.[49] They secured a Partnership Development Grant, worth $193,372, that, combined with the additional $112,060 from the industry, gave the research team significant resources "to identify which organizational practices best predict an inclusive and supportive workplace culture that maximizes organizational commitment and productivity for both men and women."[50]

To do this, the research team invited human resources professionals from 39 companies employing engineers to participate in a survey. The results showed that a minority of companies had human resources programs specifically targeting women engineers and that the tenure of women engineers was also positively related to the company's "gender diversity performance" culture. They also surveyed 269 women and men engineers and human resources personnel at their employers. Their study found that, on average, engineers from both genders "report more commitment to their organization when that company has gender inclusive policies and practices in place, where employees have positive attitudes toward these policies, and when they personally benefit from them. This pattern of results was also present for feeling valued and for intentions to leave."[51]

The very successful Partnership Development Grant led in 2017 to an impressive $2.7 million Partnership Grant addressing the issues of women in STEM. This new seven-year project titled "*Engendering Success in STEM*" (ESS) brought together a research consortium of over 30 social scientists and STEM outreach experts in addition to 22 partners in industry and education (among these CCWESTT, Engineers Canada, General Motors, ONWiE, Mining Industry Human Resources Council, Mozilla, the National Research Council, the Association of Professional Engineers and Geoscientists of British Columbia, as well as several universities). The group was devoted to "testing the long-term efficacy of interventions that harness the power of positive social interactions to mitigate subtle gender bias."[52] To foster women's inclusion and success in STEM, the consortium wanted to better understand the various cultural biases at work in these environments and combat those that stood in the way of creating inclusive cultures. To reduce the effects of implicit gender bias, the partnership oversaw four distinct projects targeting obstacles that

---

[49] "Engendering Engineering Success", http://wwest.mech.ubc.ca/ees/about/; Elizabeth Croft, "Engendering Engineering Success", *Canadian Society for Mechanical Engineering (CSME) Bulletin*, Spring 2017, accessed June 1, 2019, http://csme-scgm.ca/sites/default/files/BULLETIN-SPRING-2017.pdf, p. 7–8 and *Canadian Society for Mechanical Engineering Bulletin (CSME)*, Spring 2017, accessed June 1, 2019, http://csme-scgm.ca/sites/default/files/BULLETIN-SPRING-2017.pdf, p. 9.

[50] Elizabeth Croft, *Final Report: NSERC Chair for Women in Science and Engineering BC and Yukon Region (2010–2015)*, Vancouver, University of British Columbia, 2015, p. 8 and "Disclosure for SSHRC's grants", January 2013, accessed June 10, 2019, http://www.sshrc-crsh.gc.ca/transparency-transparence/disclosure-divulgation/grants-subventions/2013/jan_2013.pdf

[51] "Engendering Engineer Success", *CSME Bulletin*, Spring 2017, accessed June 1, 2019, http://csme-scgm.ca/sites/default/files/BULLETIN-SPRING-2017.pdf

[52] "About the consortium", date unknown, accessed June 20, 2019, http://successinstem.ca/about/

are unique to each step along the way from early education to industry: Project CLIMB (childhood), Project PRISM (adolescence), Project SINC (university), and Project RISE (industry).[53]

The Fourth Generation West Coast Chair: Lesley Shannon (2015–2020)

Following Elizabeth's decision not to pursue a second term, Lesley Shannon took the position of NSERC/Simba Technologies Inc. Chair for Women in Engineering and Science for the British Columbia Yukon region in 2015. Lesley, an Associate Professor in the School of Science and Engineering at Simon Fraser University (SFU) for nine years at the time of her appointment, studied computer systems design. Her primary area of research was on nontraditional computing system design such as reconfigurable computing, heterogeneous computing, and application-specific architectures, with applications in a wide range of areas including robotics, machine learning, aerospace and biomedical systems, multimedia applications, and cloud computing.[54] Since her arrival at SFU, Lesley has been a passionate proponent of gender diversity in STEM. Starting in 2006, she served as

[53] "About the consortium", date unknown, accessed June 20, 2019, http://successinstem.ca/about/

[54] "Dr. Lesley Shannon (2005)", date unknown, accessed June 24, 2019, https://www.obrienfoundation.ca/en/o-brien-fellows/where-are-they-now/dr-lesley-shannon-2005; "Chairholder Profile—Lesley Shannon", May 20, 2014, accessed June 24, 2019, www.nserc-crsng.gc.ca/Chairholders-TitulairesDeChaire/Chairholder-Titulaire_eng.asp?pid=914; "Lesley Shannon named NSERC Chair for Women in Science and Engineering", May 22, 2015, accessed June 24, 2019, https://www.sfu.ca/fas/news-and-outreach/years/2015/lesley-shannon-named-nserc-chair-for-women-in-science-and-engine.html and "Simba Technologies Joins with Simon Fraser University and the Natural Sciences and Engineering Research Council of Canada to Fund New Chair for Women in Science and Engineering for BC and Yukon", May 22, 2015, accessed June 24, 2019, https://www.simba.com/news/simba-technologies-joins-simon-fraser-university-natural-sciences-engineering-research-council-canada-fund-new-chair-women-science-engineering-bc-yukon/

faculty advisor for the university's Women in Engineering group, a body devoted to encourage and support the excellence of women who pursue engineering at SFU. During her six years as faculty advisor, she helped to establish outreach and mentoring programs for current and potential engineering students and was able to double the women student enrollment numbers from eight to 19 percent. Since 2013, she has also been acting as conference Chair for Elizabeth Croft's Chair's *Creating Connections*, a two-day event bringing together students, researchers, and industry leaders of all genders and backgrounds to discuss the future of women in STEM. Thus, when SFU's dean of the Faculty of Applied Sciences suggested that she should apply to the Chair program, it seemed to her like a natural step to take.[55]

Lesley's program was affected by various elements. First, the influence of Elizabeth's own project can be seen in Lesley's work. Since her involvement in *Creating Connections* in 2013, Elizabeth has been a mentor to her. Additionally, the fact that Elizabeth stayed in her position as Chair for another six months after Lesley stepped up greatly facilitated the transition. This probably explained why Lesley was one of the first West Coast Chairs to keep some of the branding from a previous Chair. Not only did she decide to keep the WWEST branding, but she also continued some of Elizabeth's activities. Second, her region's geography and demography also had a strong influence on her program. Former West Coast Chairs struggled to have a strong impact in communities outside the lower mainland of British Columbia. Accordingly, Yukon, with its low population density, had always been a region that was hard to reach. Lesley set out to build projects that would answer the needs of Yukon's population while also building an uninterrupted connection with the WWEST program. Lesley's agenda was also helped by the Government of Yukon's decision in 2015 to convert Yukon College into Yukon University. Without institutions with the ability to grant degrees in the territory, the government asked the Yukon College Board of Governors and staff members in 2014 to inform and guide the transition from college to university. This provided an impetus to cultivate stronger connections between the West Coast Chair and the soon-to-be university.[56]

[55] "Lesley Shannon", date unknown, accessed June 24, 2019, http://www2.ensc.sfu.ca/~lshannon/; "Who is WiE?", May 14, 2019, accessed June 24, 2019, https://www.sfu.ca/clubs/wie.html; "Lesley Shannon", date unknown, accessed June 24, 2019, http://successinstem.ca/team/lesley-shannon/; "Dual-Ing for Success: Making a Two Career Household Work", May 2, 2013, accessed June 24, 2019, http://cc-mech.sites.olt.ubc.ca/schedule/dual-ing-for-success-making-a-two-career-household-work/ and Lesley Shannon, Interview with Hannah Elizabeth Young, telephone interview, May 31, 2018.

[56] Lesley Shannon, Interview with Hannah Elizabeth Young, telephone interview, May 31, 2018, and Government of Yukon Press Release, "Yukon College to become Yukon University", November 4, 2015, accessed June 29, 2019, http://www.gov.yk.ca/news/15-338.html

Lesley's program was divided into four sets of activities targeting: community, industry, academia, and research. First, highlights of the community activities included the continuation of the funding of Elizabeth's Partners Program. The new West Coast Chair's team also created in April 2017 a podcast series, the "*Best of the WWEST*," in which women in STEM were interviewed about their lives and careers and experts answered questions about issues impacting diversity. The podcast series replaced an originally planned webinar series. Taking into consideration the bandwidth problems experienced by Yukon and other remote areas and the rising popularity of podcasts as well, Lesley's team decided that this medium would be more practical and scalable. The goal of the podcast series was to provide young girls with role models and discuss a range of job possibilities in these fields. Episodes were released at least twice per month. Since its launch in April 2017, the podcast has 111 episodes published which were available through Overcast, iTunes/Apple Podcasts, Stitcher, Spotify, and iHeartRadio. Concerns of Internet accessibility were also at the origin of a second project, the creation of free open-source desktop applications. The goal was to enable outreach groups "to engage youth in computing science activities regardless of their access to internet and up-to-date computers."[57] Lesley's team directly funded a first application, called Tune Twister, which provided an opportunity for students in Grades 4–8 to play with music and learn about the science of sound as well as a second application to let them have fun with image manipulation. To expand the project, they partnered with *Science AL!VE* and *Let's Talk Science* at SFU to secure an NSERC PromoScience grant. With a funding of $70,468 over three years, they started to develop a third application about machine learning. The grant was also intended to fund outreach groups to bring the applications to aboriginal, rural, and inner-city youth and to develop a teacher's toolkit with lesson plans to support British Columbia's new computing curriculum.[58]

---

[57] Lesley Shannon, *24 Month Report: NSERC Chair for Women in Science and Engineering BC and Yukon Region (2015–2017)*, Vancouver, Simon Fraser University, 2017, p. 3.

[58] Lesley Shannon, Interview with Hannah Elizabeth Young, telephone interview, May 31, 2018; "Westcoast Women in Engineering, Science and Technology", date unknown, accessed June 29, 2019, http://cwse-on.ca/80-national-network-team/310-lesley-shannon; "STEM Outreach Apps",

For her industry activities, Lesley's team focused on supporting diversity in STEM professions by improving the retention and promotion of women in industry organizations. One of her major initiatives, building on Elizabeth's work on the *Gender Diversity 101* project, aimed to engage industry organizations open to improving diversity among their employees: Lesley and her team provided them with resources and best practices on recruitment, retention, and hiring. They also gave the industrial partners the opportunity to participate in research regarding diversity in STEM. Additionally, they helped industry partners connect with community partners to support STEM outreach with role models from the industry. Between March 2017 and May 2018, the Industry Partners Program raised $3,000, financial contributions that eventually would be invested into other WWEST programs.[59]

For her work related to academia, Lesley focused her attention first on running *Creating Connections* biennial conferences for STEM professionals. The conference was started as a symposium in 2007 by Elizabeth Croft and one of her students. "*Building Communities*," which is now the *Creating Connections* conference series, was a two-day event for 120 women in engineering in British Columbia from all levels of the profession. As of 2020, the event had grown into six biennial conferences, gaining audiences of hundreds of STEM professionals, academics, and students. Covering topics on gender diversity in STEM, it was also intended to provide a venue for meaningful dialogue about diversity in these fields. Through this conference series, Lesley set the stage for individuals and organizations to build their own capacity and, in the long term, participate in changing the culture in these fields. In six years, the attendance increased by 50%. Not only had the numbers grown during these years, but the event had become more inclusive by promoting the participation of other underrepresented groups in STEM such as people of color, nonbinary people, Indigenous, and people of different abilities.[60]

---

date unknown, accessed June 29, 2019, http://www.sfu.ca/wwest/projects/STEM-outreach-apps.html; Lesley Shannon, *24 Month Report: NSERC Chair for Women in Science and Engineering BC and Yukon Region (2015–2017)*, Vancouver, Simon Fraser University, 2017, p. 2–6; Lesley Shannon, *36 Month Report: NSERC Chair for Women in Science and Engineering BC and Yukon Region (2015–2018)*, Vancouver, Simon Fraser University, 2018, p. 2, accessed June 29, 2019, http://www.sfu.ca/content/dam/sfu/wwest/images/Resources/Public%20Report%20WEB%20pages%20Version%208.pdf and "Best of the WWEST: Inspiring Women with Real Stories", June 18, 2019, accessed June 29, 2019, http://www.sfu.ca/wwest/projects/best-of-the-WWEST.html

[59] Lesley Shannon, *24 Month Report: NSERC Chair for Women in Science and Engineering BC and Yukon Region (2015–2017)*, Vancouver, Simon Fraser University, 2017, p. 4 and Lesley Shannon, *36 Month Report: NSERC Chair for Women in Science and Engineering BC and Yukon Region (2015–2018)*, Vancouver, Simon Fraser University, 2018, p. 2–3, accessed June 29, 2019, http://www.sfu.ca/content/dam/sfu/wwest/images/Resources/Public%20Report%20WEB%20pages%20Version%208.pdf

[60] Lesley Shannon, Interview with Hannah Elizabeth Young, telephone interview, May 31, 2018; Anne Millar and Mary Wells, "Elizabeth Croft Brings Women in STEM Together", *CIM Magazine*, September 14, 2017, accessed May 15, 2019, http://magazine.cim.org/en/voices/elizabeth-croft/; Anja Lanz, Elizabeth Croft, Jennifer Pelletier, Bernadette Currie, and Erin Biddlecombe, "Building Communities Symposium '07: Designing a Networking Event for Women in Engineering in BC",

To better serve the entire geography of the British Columbia and Yukon region, Lesley decided to create the WWEST Associate Chairs Program. She targeted up to three satellite programs at NSERC-eligible institutions in her region with $10,000 invested to support activities in each location. A one-year term with option for renewal, the Associate Chair Program required a minimum of $10,000 in matching funding each year. Building on the momentum created by the Government of Yukon and NSERC's decision to give additional money to the Chair program, the West Coast Chair announced in January 2018 that the first Associate Chair would be held by Alison Anderson, Technology Innovation Officer at Yukon College. In support of her projects to promote women in STEM in the Yukon, Alison was able to collect a total of $24,500 from ten Yukon organizations. In addition to her participation in the outreach program started by Lesley, the new Associate Chair planned to establish a WWEST Partners Program in the Yukon and provide engagement and support to STEM champions in Yukon communities. She also contributed to several episodes of the "Best of the WWEST" podcast and published a report in December 2018 presenting the first-ever snapshot of the state of women and girls in STEM in Yukon.[61]

In the research arena, Lesley quickly became involved in Elizabeth's *Engendering Engineering Success* project. As a co-investigator and the Industry Liaison for the group, Lesley secured two industry participants in the project: Mozilla and Magnitude: Simba Technologies. A second research project launched in 2017 administered awareness and self-efficacy surveys. The surveys were intended to analyze the short- and long-term impact of the *Creating Connections* conference series on the attendees. Lesley's team used the *Creating Connections 5.0* conference in May 2017 to determine whether they "can effect a measurable change in someone's awareness of the value of gender diversity in the technical workplace" and

paper presented to the *12th CCWESTT Conference*, Guelph, Ontario, May 29–31, 2008, accessed June 29, 2019, https://www.cs.ubc.ca/~condon/jadeproject/CCWESTT-BuildingCommunities Symposium.pdf; Lesley Shannon, *24 Month Report: NSERC Chair for Women in Science and Engineering BC and Yukon Region (2015–2017)*, Vancouver, Simon Fraser University, 2017, p. 4; Lesley Shannon, *36 Month Report: NSERC Chair for Women in Science and Engineering BC and Yukon Region (2015–2018)*, Vancouver, Simon Fraser University, 2018, p. 2–3, accessed June 29, 2019, http://www.sfu.ca/content/dam/sfu/wwest/images/Resources/Public%20Report%20 WEB%20pages%20Version%208.pdf and WWEST, *Creating Connections 6.0 Event Report*, Vancouver, Simon Fraser University, 2019, p. 1, accessed June 29, 2019, http://www.sfu.ca/content/dam/sfu/wwest/images/cc6/CC%206.0%20Wrap%20Report.pdf

[61] Lesley Shannon, Interview with Hannah Elizabeth Young, telephone interview, May 31, 2018; Lesley Shannon, *24 Month Report: NSERC Chair for Women in Science and Engineering BC and Yukon Region (2015–2017)*, Vancouver, Simon Fraser University, 2017, p. 1; Lesley Shannon, *36 Month Report: NSERC Chair for Women in Science and Engineering BC and Yukon Region (2015–2018)*, Vancouver, Simon Fraser University, 2018, p. 2; "Yukon College's Alison Anderson selected to champion STEM diversity in territory", January 24, 2018, accessed June 29, 2019, https://www.sfu.ca/wwest/WWEST_blog/yukon-college_s-alison-anderson-selected-to-champion-stem-divers.html; "WWEST Associate Chairs Program", date unknown, accessed June 29, 2019, https://www.sfu.ca/wwest/projects/associate-chairs-program.html and "WWEST at Yukon College releases Current State of Women in STEM in Yukon report", December 6, 2018, accessed June 29, 2019, https://www.yukoncollege.yk.ca/news/201812/wwest-yukon-college-releases-current-state-women-stem-yukon-report

"how effective this type of activity can be at improving occupational self-efficacy - a strong predictor of career persistence."[62] In terms of short-term impacts, the result of the surveys indicated that the conference brought significant positive change for participant awareness of the value of gender diversity but no significant change in change in their self-efficacy. Post-event surveys were also sent to all the attendees following the *Creating Connections 5.0*; 84% of them rated their experience as an 8 or higher out of 10. Furthermore, 95% of attendees would recommend the event to a colleague or friend. The WWEST team repeated the experience in 2019 for the *Creating Connections 6.0* conference with slightly better ratings (8.5 out of 10 and 97%, respectively). In terms of accessibility of the conference, an average rating of 8.8 out of 10 was recorded.[63]

Struggles Faced by the Chairs

Like the generations that preceded them, the third and fourth generations of Chairholders were anxious to increase the participation of women in science and engineering, and to provide role models for women active in, and considering, careers in these fields. They threw themselves into their work, using their predecessors' examples and their own motivations and tremendous energy to tackle problems they had lived, witnessed, and learned about. Like their predecessors, they were challenged by the heavy workload that this position entailed. The broad mandate of the Chair and the vast range of responsibilities meant that more than 50% of their time was spent on Chair-related activities (even if the NSERC program guidelines stipulated that the Chair should take up 50% of their time). Not only did the Chairs have to manage the search for funding, the hiring and training of staff, carrying out all the components they had planned for (usually outreach, community building, program development, and implementation as well as research), but they also had to undertake the direction of two very distinct research programs (one for the Chair and one for their "regular" science or engineering research). It is important to note that to hold a CWSE Chair the candidates have to have an active NSERC Discovery grant for their "regular" science or engineering research and remain at

[62] *Creating Connections 5.0 Event Report*, Vancouver, Simon Fraser University, 2018, p. 4, accessed June 29, 2019, https://www.sfu.ca/content/dam/sfu/wwest/images/cc5/CC5.0_Wrap%20Report_Design_V6_WEB.pdf

[63] Lesley Shannon, *24 Month Report: NSERC Chair for Women in Science and Engineering BC and Yukon Region (2015–2017)*, Vancouver, Simon Fraser University, 2017, p. 4–5; Lesley Shannon, *36 Month Report: NSERC Chair for Women in Science and Engineering BC and Yukon Region (2015–2018)*, Vancouver, Simon Fraser University, 2018, p. 3 and *Creating Connections 5.0 Event Report*, Vancouver, Simon Fraser University, 2018, p. 4, accessed June 29, 2019, https://www.sfu.ca/content/dam/sfu/wwest/images/cc5/CC5.0_Wrap%20Report_Design_V6_WEB.pdf and *Creating Connections 6.0 Event Report*, Vancouver, Simon Fraser University, 2019, p. 4, accessed June 29, 2019, http://www.sfu.ca/content/dam/sfu/wwest/images/cc6/CC%206.0%20Wrap%20Report.pdf

the top of their field to do so. For some, the Chair responsibilities affected their ability to pursue their regular research in their own (science or engineering) area of expertise. For others, the burden of juggling all these things at the same time was overloading and stressful.[64]

This problem was brought to the attention of NSERC by the second generation of Chairholders: in particular Valerie Davidson and Nadia Ghazzali pushed hard for additional funding for one postdoctoral fellow position for each Chair. The pressure on university scientists to publish research papers in great quantity was—and still is—relentless and an unavoidable part of the academic environment. Since the research done for the Chair was often not taken into account for promotion purposes in their department or school, the Chairholders needed to find a way to continue to publish in the researcher's primary expertise field. Having a postdoctoral fellow who could be more independent and provide training to graduate students while working on the Chairholders' science or engineering research projects would give them some support to keep up their research profile. NSERC agreed in 2010 to add support for a postdoctoral fellow in the program description. The funding would again follow a matching model, with NSERC providing $25,000 and requiring the applicant's university to supply at least $20,000. This support was enormously helpful to the third and subsequent generations of Chairs. The challenge to uphold their research output while performing all the Chair's activities was so grueling that at least one of the Chairs admitted that she would not have obtained her tenure without the help provided by her postdoctoral fellow. For her, the postdoctoral position was instrumental to her productivity in her primary expertise field. Yet, despite the introduction of this support, at least one of the Chairholders had problems filling the position partly due to the low salary rate. Indeed, the postdoctoral fellowship provided with the Chair had an annual stipend of only $45,000 per year while most of these positions in the field were offered at $60,000–$75,000 per year. For computer science, the rates could be significantly higher.[65]

The lack of support from colleagues was another recurring complaint among the Chairholders. Their engagement to actively promote women in science and engineering was still met with indifference in some parts of the academic sector. Of course, some of their peers thought the whole Chair program was a wonderful idea, and they really were working hand in hand with the Chairholders. For example, the Ontario Chair, Catherine Mavriplis, remarked that her own dean was very interested in what she did and attended most of her events. Still, others did not understand why

[64] Lesley Shannon, Interview with Hannah Elizabeth Young, telephone interview, May 31, 2018, and Ève Langelier, Interview with Hannah Elizabeth Young, telephone interview, May 31, 2018.

[65] Nadia Ghazzali, Interview with Hannah Elizabeth Young, telephone interview, July 3, 2018; Valerie Davidson, Interview with Hannah Elizabeth Young, telephone interview, February 3, 2018; Tamara Franz-Odendaal, Interview with Hannah Elizabeth Young, telephone interview, June 1, 2018; Lesley Shannon, Interview with Hannah Elizabeth Young, telephone interview, May 31, 2018; Ève Langelier, Interview with Hannah Elizabeth Young, telephone interview, May 31, 2018; and Annemieke Farenhorst, Interview with Hannah Elizabeth Young, telephone interview, June 6, 2018.

they were involved in this while others thought it was a complete waste of time: it was just a bunch of women complaining or it was just doing women's work. These colleagues just did not see the value of diversity and inclusion or did not see the value of the research done behind this effort because it was mostly humanities and social sciences research. Many of them saw this question as recruiting women at the expense of excellence, as if diversity, inclusion, and excellence cannot go together. The Chairholders had to battle the academic culture and most often had to face it on their own as five isolated leaders spread across the country. In contrast, in the United States, NSF's programs to support women in STEM had changed dramatically through the years: by 2001, they had pivoted to an "institutional transformation" model to change the culture of the whole academic institution, with the introduction of the ADVANCE program. For the large ADVANCE grants, the principal investigator had to be a provost, a dean, or someone in the higher administration of the applicant institution.[66]

A final recurring challenge for the Chairholders was the difficulty to secure industrial or private support for their Chair. For some of them, scientists in particular, it was really hard to engage industry because their own expertise was outside the more lucrative engineering or pharmaceutical fields. As a result, they dedicated much of their time to fundraising at the expense of their own work for the promotion of women in science and engineering. For others, the challenge came from their geographical situation. If the Chair was allocated to a university in a more rural region, chances were that there was no big company in the area with the capacity or the willingness to finance such an undertaking. The Chairholder would then have to drive to the larger cities in their region to engage big companies with financial capacity. Most universities also have strict rules on how sponsorships can be obtained and who can approach the industry for support. Some Chairs were not able to approach companies directly and were prevented by their universities' advancement office rules to approach certain companies. These rules are in place to follow strategic plans determining who can and when to approach big donors. As a unique and poorly understood Chair, it was often difficult for Chairholders (and CWSE program applicants) to navigate the fundraising arena.[67]

***

[66] Lesley Shannon, Interview with Hannah Elizabeth Young, telephone interview, May 31, 2018; Catherine Mavriplis, Interview with Hannah Elizabeth Young, August 9, 2018; Tamara Franz-Odendaal, Interview with Hannah Elizabeth Young, telephone interview, June 1, 2018; and Annemieke Farenhorst, Interview with Hannah Elizabeth Young, telephone interview, June 6, 2018.

[67] Ève Langelier, Interview with Hannah Elizabeth Young, telephone interview, May 31, 2018; Ève Langelier, *Rapport d'étape de 24 mois (mai 2015 à avril 2017)*, Sherbrooke, Université de Sherbrooke, January 2018, p. 8; Catherine Mavriplis, Interview with Hannah Elizabeth Young, August 9, 2018; and Tamara Franz-Odendaal, Interview with Hannah Elizabeth Young, telephone interview, June 1, 2018.

Like the previous Chairs for Women in Science and Engineering, the six women of the third generation (and the fourth for BC and Yukon) worked together to increase the participation of women in science and engineering and provide role models for women active in, and considering, careers in these fields. Through their collective efforts, they reached out to more than 180,000 girls and their parents, women, industry professionals, as well as academics and teachers to promote careers in science and engineering and better inclusion in the workplace. They shared proven practices with more than 70 committees and through more than 400 lectures, panels, courses, workshops, and conferences. Together, they were able to shine the spotlight on this critical issue by publishing more than 200 academic publications and by giving more than 175 interviews. They built significant momentum by securing more than 15 major grants and leveraged more than $9 million beyond the original NSERC and matching funds awarded for their Chairs. And finally, they built a strong network of more than 70 companies, 170 organizations, and six international partnerships.[68]

Their impact reached far beyond these direct interactions. The regional Chairs were also engaged with institutional hiring and diversity committees, industry- and sector-specific gender-equality drives, professional association initiatives, and the collection of efforts that these stakeholders champion. These time-consuming activities were also important to changing the culture of the work environment in science and engineering fields and encouraging female students and women to consider, and continue, a career in these domains. Although the drive to integrate more women in STEM fields is not finished, the third generation of Chairs for Women in Science and Engineering once again broadened the scope of engagement, particularly through novel industry interactions, sustained policy consultations, and the building of new networks through large in-person workshops and conferences as well as novel social media avenues that arose during their terms.[69]

[68] Chairs for Women in Science and Engineering National Network, *Coast to Coast update 2016*, Ottawa, Réseau National Network, 2016, p. 14–15 and Chairs for Women in Science and Engineering National Network, *Coast to Coast update 2018*, Ottawa, Réseau National Network, 2018, p. 13–14.

[69] Jane Whynot, Catherine Mavriplis, Annemieke Farenhorst, Ève Langelier, Tamara Franz-Odendaal, and Lesley Shannon, "Knitting Theory in STEM Performance Stories: Experiences in Developing a Performance Framework", *Canadian Journal of Program Evaluation*, vol. 33, no 3 (2019), p. 360.

Chapter 5
Epilogue

For 30 years, a select group of women scientists and engineers have passionately devoted themselves to advancing girls and women in the study and practice of science and engineering in Canada and beyond through their tenure as Natural Sciences and Engineering Research Council of Canada (NSERC) Chairs for Women in Science and Engineering. Their mandates have been very large, unwieldy at times, and yet they poured their energies into the job, attempting to fill every corner of the wide space conferred to them: often seen as a three-dimensional space with one axis being the study/career level from young girls to executive leader women, a second axis being the range of disciplines from biological or soil sciences to the varied fields of engineering, and the third axis the wide geographical one of their region as well as contributing to the national and international arenas.

In this book, we have merely scratched the surface of the wide and deep array of activities that the 16 Chairs from 1989 to 2019 undertook. We invite you to explore their publications and other materials (in the form of websites, videos, media interviews, and much more) either online or in the archives they have created or will create, through their host universities or the national *Canadian Archive of Women in STEM* (at the University of Ottawa). Since 2020, three new Chairs have been appointed: Laleh Behjat, a professor of electrical and computer engineering at the University of Calgary in the Prairie region, in 2020; Shohini Ghose, a professor of Physics at Laurier University in the Ontario region, in 2020; and Jennifer Jakobi, a professor of exercise science at the University of British Columbia Okanagan campus, in 2021 (See Fig. 5.1). Since the NSERC Chair for Women in Science and Engineering (CWSE) program allowed for five-year terms with renewals of three years (up until the second generation) and/or five years later, with many Chairs opting to renew and some moving on to other leadership opportunities, the onboarding of new Chairs in the different regions did not always line up. A timeline of the Chairs is given in Fig. 5.2.

C. D'Amours et al., *The Chairs for Women in Science and Engineering Program*, Women in Engineering and Science, https://doi.org/10.1007/978-3-031-54085-1_5

Fig. 5.1 The latest NSERC Chairs for Women in Science and Engineering: from left to right, Prof. Laleh Behjat (since 2020), Prof. Shohini Ghose (since 2020), and Prof. Jennifer Jakobi (since 2021)

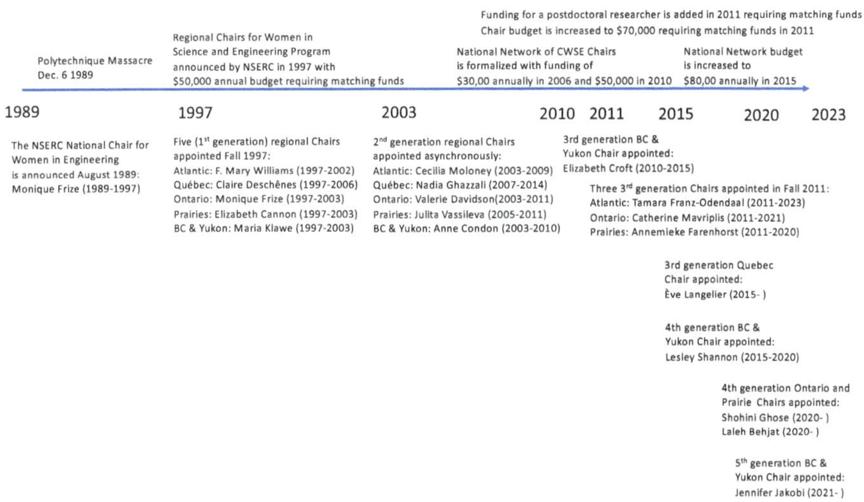

Fig. 5.2 A timeline of the NSERC Chairs for Women in Science and Engineering Program

While the Chairs were not privy to the selection process, they appreciated the balance of scientists and engineers that was maintained in the course of the program. Overall, there have been eleven engineers and eight scientists in CWSE positions. Together, they shared wider concerns of the advancement and treatment of women in science and engineering, while learning about each other's disciplines and challenges, practices and connections to industry and other stakeholders. The idea to have five regional Chairs rather than one from 1997 onward turned out to be most beneficial for the work to get done in a more equitable manner across this wide country. While the workload was still much too high for only five women, who were expected to spend only 50% of their time on the Chair, the shared sense of responsibility and intimacy of the group of five strengthened their efforts and provided a very supportive and collegial community that the Chairs treasured during their tenure and far beyond. The uniqueness of the five positions across the country and the naming of "NSERC Chair" also lent a certain prestige to the program which attracted attention: Chairs are often sought after for media interviews, especially on the 6th of December, the anniversary of the Polytechnique massacre. In industry and abroad, the Chair title is considered to be prestigious. The title alone generates a large number of invitations for keynote speeches, visits to institutions, participation on large interdisciplinary grant proposal teams, and requests for policy consultations.

Most Chairs depended on a strong "right-hand" woman (or man) to assist them and manage the multiple aspects of their Chair program. These are listed in Table 5.1 along with other key Chair information (see also Fig. 5.3). They were most often women who had training in science or engineering and were looking to support women through outreach or skills development activities. Since many women eventually leave the academic or standard career path in these fields, there was a ready pool of people interested in working in this arena. Most had formidable organizational skills, and their contributions to the Chair programs were significant, not only in running events, maintaining budgets and metrics, but also in providing creative and intellectual content to program development. For example, Carolyn Emerson is co-author of the book Becoming Leaders with Mary Williams and single-handedly produced the second edition of the book published by the American Society of Civil Engineers Press.

Thanks to the persistent negotiation of the second-generation Chairs, Valerie Davidson and Julita Vassileva in particular, NSERC agreed to fund the National Network of CWSE Chairs with an annual budget of $30,000 for five years. The budget was to facilitate biennial face-to-face and teleconference meetings of the Chairs and any joint projects they undertook. Valerie Davidson, Ontario Chair, chaired the network from 2006 to 2011, when Elizabeth Croft, BC & Yukon Chair, took over. Chairing the network came with the financial responsibility of the budget, with its expenses, transfers, and reimbursements, as the funds were allocated to that person's institution. The proposal for an increased budget of $50,000 per year was accepted for the second network term of 2010–2015. Elizabeth led the collaborative drafting of the annual and final reports and the proposal among all five Chairs for the next five years in 2014. The network agreed to ask for an increase in funding at that time. A budget proposal for $80,000 per year was accepted as the network had increased the regularity of their meetings, both by teleconference and in person, to nearly monthly and had developed joint national research projects and regular national workshop activities. The network also supported third-party groups organizing activities benefiting women in science and engineering, such as sponsorship of national conferences in support of women in specific fields or generally in science and engineering (e.g., now known as the Margaret-Ann Armour National Conference Grant program) and leadership workshops held in different regions of the country. In late 2015, as Elizabeth transitioned out of the Chair program, Catherine Mavriplis, Ontario Chair, took over the management of the National Network and the newly increased budget of $80,000. She then led the collaborative drafting of the final report and the proposal among all five Chairs for the next

Table 5.1 NSERC CWSE Chair program names, dates, sponsors, Chairholders, and primary personnel

Chair program	Sponsor(s)	Chairholder Institution Discipline	Primary assistants/ managers
National Chair (1989–1997) The Northern Telecom/ NSERC Women in Engineering Chair From 1995: The Nortel/NSERC Women in Engineering Chair See Chap. 1	Northern Telecom	Monique Frize University of New Brunswick Electrical engineer	Jane McGinn-Giberson Gisèle Levesque
Atlantic Chair (1997–2002) The NSERC/Petro-Canada Chair for Women in Science and Engineering, Atlantic Region First Generation See Chap. 2	Petro-Canada	F. Mary Williams Memorial University of Newfoundland Ocean engineer	Carolyn Emerson
Quebec Chair (1997–2006) La Chaire CRSNG/ Alcan pour les femmes en sciences et génie au Québec First Generation See Chap. 2	Alcan	Claire Deschênes Université Laval Mechanical engineer	Martine Foisy Maryse Fortin Judith Sévigny
Ontario Chair (1997–2003) The Nortel/NSERC Women in Science and Engineering Chair First Generation See Chap. 2	Nortel Networks	Monique Frize University of Ottawa and Carleton University Electrical engineer	Several students were hired part-time
Prairie Chair (1997–2004) The NSERC/Petro-Canada Chair for Women in Science and Engineering (Prairie Region) First Generation See Chap. 2	Petro-Canada	Elizabeth Cannon University of Calgary Geomatics engineer	Marcia Inch

(continued)

Table 5.1 (continued)

Chair program	Sponsor(s)	Chairholder Institution Discipline	Primary assistants/ managers
British Columbia &Yukon Chair (1997–2003) The IBM/NSERC Chair for Women in Science and Engineering First Generation See Chap. 2	IBM Canada	Maria Klawe University of British Columbia Computer scientist	Michele Ng
Atlantic Chair (2003–2008) The NSERC/Petro-Canada Chair for Women in Science and Engineering, Atlantic Region Second Generation See Chap. 3	Petro-Canada	Cecilia Moloney Memorial University of Newfoundland Electrical engineer	Caroline Koenig
Quebec Chair (2007–2014) Chaire CRSNG-Industrielle Alliance pour les femmes en sciences et en génie au Québec Second Generation See Chap. 3	Industrielle Alliance (2007–2012) Fondation de l'UQTR (2012–2014)	Nadia Ghazzali Université Laval (2007–2012) Université du Québec à Trois Rivières (UQTR) (2012–2014) Mathematician/ Statistician	Pascale Dubé
Ontario Chair (2003–2011) CWSE-ON Second Generation See Chap. 3	Hewlett-Packard Canada (2003–2008) Research in Motion (RIM) (2008-2011)	Valerie Davidson University of Guelph Food engineer	Janice DeMoor Jenny Fender Simone Saunders Melanie Veltman
Prairie Chair (2003–2011) The Cameco NSERC Prairies Chair Women in Science and Engineering (WISE) Second Generation See Chap. 3	Cameco	Julita Vassileva University of Saskatchewan Computer scientist	Monisha Shukla Safaa Mould Wendy Sharpe Heather Truman Kristina Rabyj

(continued)

Table 5.1 (continued)

Chair program	Sponsor(s)	Chairholder Institution Discipline	Primary assistants/ managers
British Columbia &Yukon Chair (2003–2009) The Jade Project Second Generation See Chap. 3	General Motors of Canada	Anne Condon University of British Columbia Computer scientist	Michele Ng
Atlantic Chair (2011–2023) WISEAtlantic Third Generation See Chap. 4	Techsploration, Nova Scotia Community College, Skills Canada Nova Scotia, Lockheed Martin Canada, Mount Saint Vincent University	Tamara Franz-Odendaal Mount Saint Vincent University Biologist	Tanya Poulton Sally Marchand
Quebec Chair (2015–) Chaire pour les femmes en sciences et en génie (CFSG) Third Generation See Chap. 4	Fondation de l'Université de Sherbrooke, Faculté de génie de l'Université de Sherbrooke, Aéroports de Montréal, GE Aviation Bromont, Varitron Granby, Fonds de recherche du Québec – Nature et technologies, Secrétariat de la condition feminine, TM4	Ève Langelier Université de Sherbrooke Biomedical engineer	Vincent Belletête Joëlle Pelletier-Nolet Jade Brodeur
Ontario Chair (2011–2021) NSERC-Pratt & Whitney Canada Chair for Women in Science and Engineering (2011–2016) CWSE-ON (2017–2021) Third Generation See Chap. 4	Pratt & Whitney Canada (2011–2016) Leacross Foundation (2017–2020) CAE (2017–2020) Lockheed Martin Corp. of Canada (2017)	Catherine Mavriplis University of Ottawa Mechanical and aerospace engineer	Carolina Zoca Katy Pitts Angeline Blais Mays Ahmad Maha Manoubi Emilie Salinas Laura Haya Marilou Charbonneau Jennifer Thivierge Lisa Ursaki

(continued)

Table 5.1 (continued)

Chair program	Sponsor(s)	Chairholder Institution Discipline	Primary assistants/managers
Prairie Chair (2011–2020) The NSERC Chair for Women in Science and Engineering for the Prairie region Third Generation See Chap. 4	University of Manitoba, Verna J. Kirkness Education Foundation, University of Saskatchewan, Province of Manitoba (Agriculture), Manitoba Pork Council, Engineers and Geoscientists Manitoba, University of Alberta, University of Calgary University of Regina, Iris Meck Communications	Annemieke Farenhorst University of Manitoba Soil scientist	Jenna Rapai Lena Yusim Mahalia Lepage Kimberley Stefaniuk
British Columbia & Yukon Chair (2010–2015) WWEST (Westcoast Women in Engineering, Science and Technology) Third Generation See Chap. 4	UBC Faculty of Applied Science, BC Hydro, Worley Parsons Canada, Dr. Ken Spencer, Teck Resources, Stantec Consulting, Henry F. Man, Catherine Roome, Stanley Cowdell, Division for Advancement of Women in Engineering and Geoscience, Nemetz & Associates, Glotman Simpson Consulting Engineers, Karen Savage Peng, Golder Associates	Elizabeth Croft University of British Columbia Mechanical engineer	Jennifer Pelletier
British Columbia & Yukon Chair (2015–2020) WWEST Fourth Generation See Chap. 4	Orbital Technologies, Simba Technologies	Lesley Shannon Simon Fraser University Computer engineer	Danniele Livengood
Ontario Chair (2020–) NSERC Chair for Women in Science for Ontario Fourth Generation Not described in any detail in this book	Perimeter Institute for Theoretical Physics, Communitech, NXM Technologies, Conestoga College	Shohini Ghose Wlifrid Laurier University Physicist	Skye Hennessey

(continued)

Table 5.1 (continued)

Chair program	Sponsor(s)	Chairholder Institution Discipline	Primary assistants/ managers
Prairie Chair (2020–) WISE Planet Fourth Generation Not described in any detail in this book	University of Calgary, ENMAX, Canadian Utilities – ATCO, Indigenous Engineering Inclusion, Alberta Women in Science Network, Spartan Controls, Calgary Public Library, Platform Calgary, Calgary Science Centre Society	Laleh Behjat University of Calgary Electrical engineer	Jennifer van Zelm
British Columbia & Yukon Chair (2021–) WWEST Fifth Generation Not described in any detail in this book	University of British Columbia Okanagan, Westmar Advisors, Women in Consulting Engineering Vancouver	Jennifer Jakobi University of British Columbia Okanagan Exercise scientist	Rebecca McCullough

Fig. 5.3 Claire Deschênes (second from left) with her trusted assistants Judith Sévigny, Corina Oancea, and Maryse Fortin (from left to right)

five years in 2019. In the fall of 2020, Tamara Franz-Odendaal took over as manager of the National Network, and the funds were transferred from the University of Ottawa to Mount Saint Vincent University. As of 2023, Jennifer Jakobi is now managing the network budget from University of British Columbia (UBC) Okanagan campus.

The Chair programs also involved a wide number of other workers and supporters either through student internships, co-op positions, research assistantships, and volunteer positions. Often, as in the case of the WISEAtlantic Chair, these women had been participants in the outreach programs as girls and their association with the Chair continued for over ten years. This was one advantage of the extended timeline of the Chairs to ten and, in some cases, longer than ten years (due to sabbatical leaves or requests from NSERC to stay on through the onboarding process of the next Chair in the region).

The Chairs appreciated the strong support of the many NSERC CWSE program managers over the years. They found them attentive to their concerns and always willing to work with them in resolving problems and challenges. They were diligent in maintaining contact with the network of Chairs, often through half-day meetings at national conferences, such as Canadian Coalition for Women in Science, Engineering, Technology and the Trades (CCWESTT). They attended some Chair events as their travel schedule would allow, or in Ottawa when appropriate. They were actively involved in many of the evolutionary changes of the program, always in consultation with the current Chairs. At the risk of missing any names, we list the following NSERC professionals who helped the Chairs along the way: Teresa Jurewicz, Marilyn Taylor, Isabelle Blain, Barbara Muir, Anne Alper, Monika Michalska, Sarah Overington, Rawni Sharp, Enikö Megyeri-Lawless, Serge Villemure, Jennifer Houghton, Céline Bérubé, Valerie Harbour, Natalie Weiskopf, Coralie Rochefort, Kathleen Lorenzo, Nyree St-Denis, Anne Webb, Pierre Charest, Adèle Ngi-Song, Danika Goosney, Catherine Podeszfinski, Nathalie Podeszfinski, Marie-Pierre Schippers, and Sarah Smith.

Impacts and Outcomes

Fourteen universities benefited from the hosting of a CWSE Chair with four (UBC, Laval, Calgary, and Ottawa) having more than one Chair. Over 20 large private corporations have provided matching funds to make the Chairs possible, with several other smaller entities and governmental or philanthropic partners. Possibly the greatest impact was on the number of people reached with programming and activities to empower and motivate girls and women to study, pursue careers, and advance in science and engineering. For example, in the first generation, Elizabeth Cannon reported leading 222 events under her Chair program and reaching 11,395 participants in the period 1997–2001. Valerie Davidson, in the second generation, reported direct interactions with 36,000 students, 4,400 teachers, 2,500 university students and 6,200 professionals, for a total of 49,100 over the period 2003–2011. The growth in Valerie's numbers, displayed in Fig. 5.4, illustrates the benefits of extending a Chair to a second term. Once the momentum is going and the networks have been built, it becomes easier to reach people and deliver programming.

The third generation of Chairs reported their numbers collectively in the Coast-to-Coast Update brochure biennially, as shown in Fig. 5.5 for the period 2015–2019,

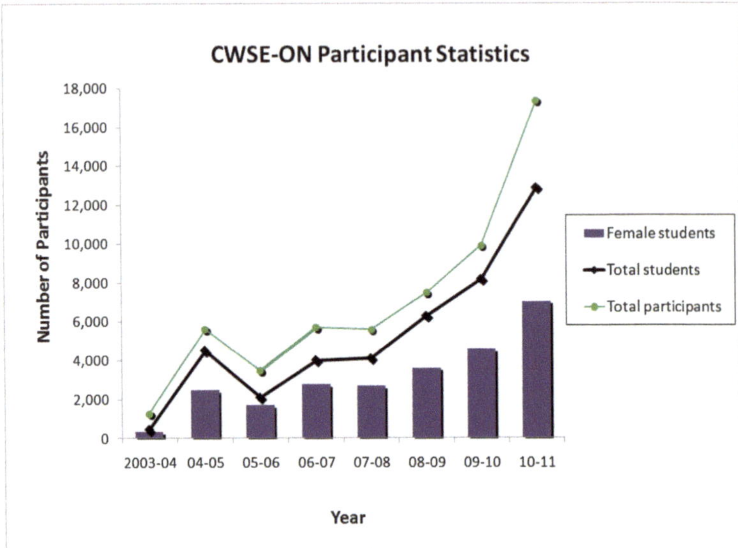

Fig. 5.4 Participation in Valerie Davidson's CWSE-ON Chair grew steadily as she extended her Chair to a second term

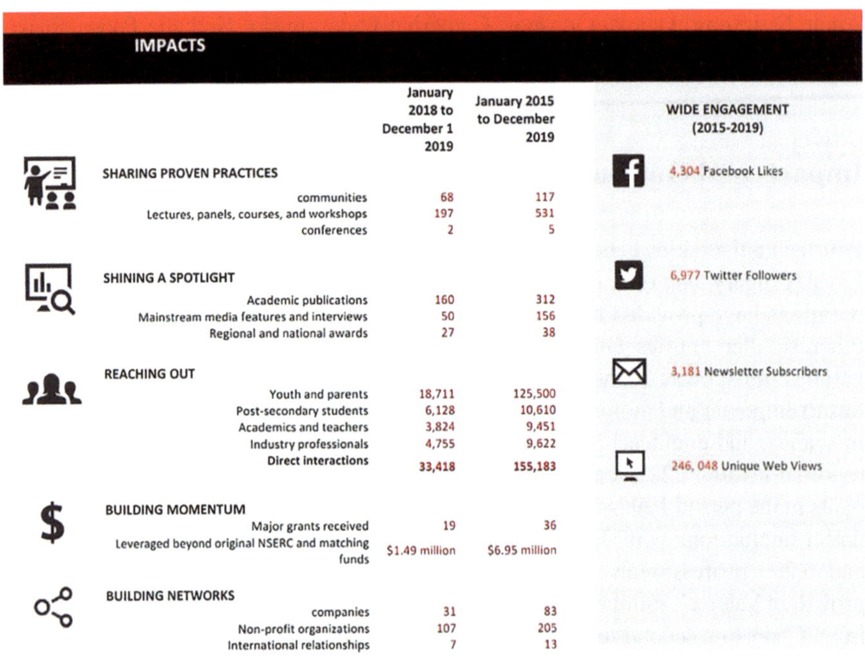
IMPACTS

	January 2018 to December 1 2019	January 2015 to December 2019
SHARING PROVEN PRACTICES		
communities	68	117
Lectures, panels, courses, and workshops	197	531
conferences	2	5
SHINING A SPOTLIGHT		
Academic publications	160	312
Mainstream media features and interviews	50	156
Regional and national awards	27	38
REACHING OUT		
Youth and parents	18,711	125,500
Post-secondary students	6,128	10,610
Academics and teachers	3,824	9,451
Industry professionals	4,755	9,622
Direct interactions	33,418	155,183
BUILDING MOMENTUM		
Major grants received	19	36
Leveraged beyond original NSERC and matching funds	$1.49 million	$6.95 million
BUILDING NETWORKS		
companies	31	83
Non-profit organizations	107	205
International relationships	7	13

WIDE ENGAGEMENT (2015-2019)

4,304 Facebook Likes

6,977 Twitter Followers

3,181 Newsletter Subscribers

246, 048 Unique Web Views

Fig. 5.5 Impacts of the five CWSE programs and their National Network for the period 2015–2019 as measured by a set of metrics developed by the Chairs in 2013

where 155,183 people were involved in direct programming with at least one of the five Chairholders across Canada. The rise of social media in this period allowed the Chairs to greatly widen their reach. A conservative estimate for the number of people reached would be over half a million people over 22 years from 1997 to 2019. During the COVID pandemic, the numbers were reduced as programming could only happen virtually and activities had to be pivoted to online offerings.

In the same reporting period of 2015–2019, the National Network reported working collectively with over 300 organizations. The current (2023) list of partners is displayed in Fig. 5.6, showing a wide and varied array of partners from coast to coast.

Major outcomes for the Chair program include several long-standing outreach programs, such as *Pathmakers, Explore IT, Cybermentor, Go Eng Girl, Go Code Girl, Les filles et les sciences – un duo électrisant!* Many of these continue to operate through partner organizations such as Ontario Network of Women in Engineering (ONWiE). The creation of ONWiE and other organizations are also major outcomes of the Chairs program. AFFESTIM has been in existence since 2003 and has provided programming, publications, and resources for francophone women in Quebec mostly, but also in New Brunswick, Nova Scotia, and Ontario. Since 2005, ONWiE networked the 26 schools of engineering across Ontario to provide a uniform offering of outreach programs to attract girls to engineering. International Network of Women Engineers and Scientists (INWES) was created in 2002, networking over 250,000 women in science and engineering from support organizations across the world, from over 60 countries. INWES has hosted the International Conference of Women Engineers and Scientists (ICWES) conferences since 2002. Finally, the creation of the *Canadian Archive for Women in STEM*, housed at the University of Ottawa in partnership with Library and Archives Canada and The Canadian Institute

Fig. 5.6 Partnership map of the National Network of CWSE Chairs in 2023

Fig. 5.7 Assembled dignitaries at the establishment of the national *Canadian Archive of Women in STEM* at the University of Ottawa in 2018: from left to right, the University of Ottawa President, Jacques Frémont; Library and Archives Canada Director, Guy Berthiaume; CWSE Chairs, Monique Frize and Claire Deschênes; Édith Ducharme, student from Laval; Chief Science Advisor, Mona Nemer; Professor of History, Ruby Heap; Mélissa Fortin, student from Laval; the University of Ottawa Chief Librarian, Leslie Weir; and CWSE Chair, Catherine Mavriplis. Minister of Science, the Honorable Kirsty Duncan, was on hand for the ceremony and took the opportunity to network with graduate student women in science

of Women in Engineering and Science (CIWES), was a major outcome for the Chair program (see Fig. 5.7).

The Chairs program provided early programming for Indigenous youth to consider science and engineering studies, mostly through the Prairie Chairs but also as early as 1996 in Monique's first Chair. Julita Vassileva's Chair had a major focus on Indigenous youth with the Science Ambassador Program which reached 3,150 students over four years in ten communities. Annemieke Farenhorst's Prairie Chair grew the reach and participation of Indigenous students significantly, as seen in the impacts map of Fig. 5.8. For example, the Verna J. Kirkness Education program launched in 2012 at the University of Manitoba with the Prairie Chair with eight students and grew to eight institutions and 141 students by 2019. Annemieke's NSERC CREATE H2O program, centered on First Nations water and sanitation, recruited Indigenous graduate students as 33% of the six-year program. Other Chairs joined in taking cues from Annemieke's experiences and broadened their reach as well.

The Chair program has served as a model that has been replicated in a few spots across the country. First, Valerie Davidson convinced the Council of Ontario Deans of Engineering to create and support an ONWiE Chair (Ontario Network of Women in Engineering) which has been in existence since 2013. Designed to manage the extensive GoEng Girl outreach program and others like it, the Chair rotates between universities in Ontario, with Mary Wells leading from 2013 to 2018 at the University of Waterloo, and Kim Jones since then at McMaster University. In 2014, Goldcorp

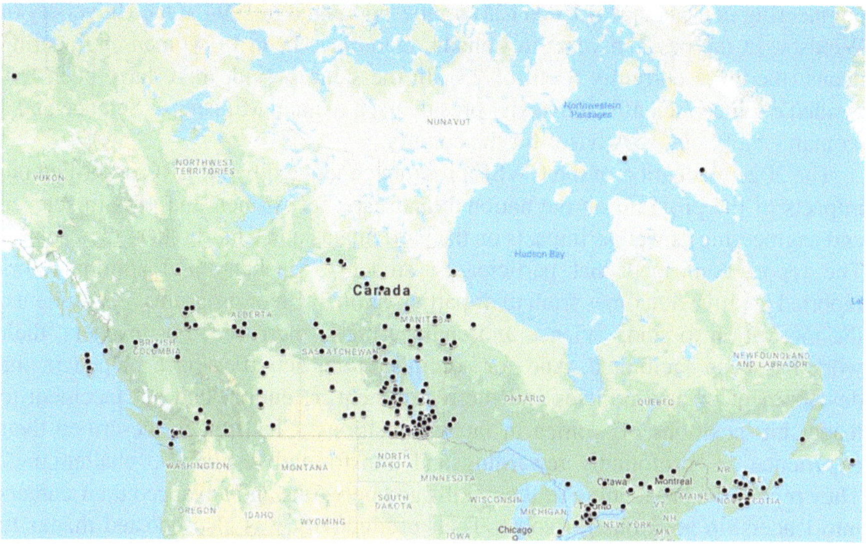

Fig 5.8 Regional impact of the CWSE Chair programs from 2015 to 2019

donated $500,000 to UBC's Faculty of Applied Science to create a named professorship for women in engineering, held by civil engineering professor Sheryl Staub-French. She is now Associate Dean for Equity, Diversity, and Inclusion. In 2020, biomechanics professor Heidi-Lynn Ploeg became the inaugural Chair for Women in Engineering at Queen's University thanks to an anonymous alumni gift of $3M. In 2022, the University of Guelph announced a $1M gift for the Doody Family Chair for Women in Engineering, matched by over $2M from the university. Civil engineering professor Jana Levison will hold the inaugural Chair for five years.

Over the period of the CWSE program, the number of women and girls in Science, Technology, Engineering and Mathematics (STEM) has indeed risen dramatically. For example, in engineering studies, 21,859 women (24.2%) were enrolled in undergraduate programs in 2020,[1] compared with 4,435 women (13%) in 1989.[2] Master's degree enrollments of 9.5% and PhD degree enrollments of 5.5% in 1989 also rose: in 2020, 27.4% of the postgraduate enrollment was by women. While Engineers Canada is encouraging universities to increase enrollments so that the target of 30% of newly licensed engineers being women may be reached by 2030, through their "30 by 30" campaign, outreach and other efforts toward systemic change remain important to reach that goal. In September 1990, women accounted for only 2.2% of the faculty members in engineering across Canada. By 2020, women made up 17.3% of the faculty members. The first woman dean of

[1] Engineers Canada, Canadian Engineers for Tomorrow: Trends in Engineering Enrolment and Degrees Awarded 2020, 2021. https://engineerscanada.ca/reports/canadian-engineers-for-tomorrow-2020

[2] Source: Enrolment and Degrees Awarded in Engineering, 1990, A report of the Canadian Engineering Human Resources Board.

engineering in Canada was Hoda ElMaraghy, appointed in 1994 at the University of Windsor. In the past ten years in Canada, there have been 11 women engineering deans (there are currently 8 out of 25). In the same period, there have been five women engineer/scientist university presidents, a woman Minister of Science and a woman Chief Science Advisor for the country.

The 2021 evaluation of the CWSE program conducted by NSERC described the impacts of the program on the national landscape for women and girls in science and engineering as well as impacts on the participants and Chairholders themselves. The report confirmed that participants, students, teachers, and administrators reported positive outcomes from their participation in Chair programs. For students, the motivation to study science and engineering is increased. For teachers, their work has been facilitated. And for administrators, activities have helped in the development of new policies and the recruitment, retention, and advancement to leadership positions of women in faculty positions. Chairholders described their experience as "profoundly rewarding and transformative, but also challenging." They reported a broadening of their skills and networks and advanced their careers into leadership positions for some. Their productivity was also boosted thanks to partnerships and, for the latter Chairs, the postdoctoral support. The evaluation process confirmed the importance of the National Network as a communication and sharing channel, allowing activities to be shared and reused across the country as well as ensuring continuity from Chair to Chair. Partnerships were also found to be "invaluable" to the Chairs, but admittedly hard to secure from industry at times.

The recommendations in the report were that:

1- "NSERC should maintain the CWSE program," noting that "the program also serves to communicate NSERC's vision and commitment related to equity, diversity and inclusion in the fields of natural sciences and engineering."
2- "NSERC should review the current set of program objectives to ensure that they more closely reflect the range of activities undertaken by the chairholders and the contribution that the program can be reasonably expected to achieve."
3- "NSERC should review the process related to transition between chairholders in each region to include overlap for knowledge sharing and continuity."[3]

How to Measure Impact

While the Chairs reported in 24-month and 48-month as well as final reports to NSERC on all the activities and research they were undertaking and what the results were, it was hard to do so in the restricted ten-page format. Many of the outcomes were intangibles like increased self-confidence and empowerment of girls and women, whether they pursued science and engineering studies and careers or not. Furthermore, it was difficult for Chairs to cover their entire geographical regions

[3] NSERC, Evaluation of the Chairs for Women in Science and Engineering Program, 2021. https://www.nserc-crsng.gc.ca/_doc/Reports-Rapports/CWSE-Eval_e.pdf

and assume that increases in enrollment numbers were entirely due to their efforts. NSERC in the meantime struggled with the desire to measure the long-term impact of the program. There was no template for the Chairs to follow, and NSERC managers approached the Chairholders at regular intervals to ask them to develop their own metrics and evaluation procedures.

Valerie Davidson wrote in her final report in 2012:

> At the meetings of the regional Chairs and NSERC staff between 2004 and 2006, there was considerable discussion of performance measures for the CWSE program. Each regional chair develops a short-term (i.e. 5- to 8-year) program with a regional focus. NSERC's long-term goal is to increase the national participation of women in science and engineering – a crucial but complex social innovation. In our discussions, we recognised the importance of evaluation but also the challenges due to time frames and resources (time and money) for evaluation.
>
> In 2006, I developed program logic model for the Ontario program as a first step. This was about the mid-point of the first term (2003-2008) so the logic model represented some degree of hindsight in terms of program evaluation. The logic model [shown in Fig. 5.9] attempted to define short- and intermediate-term impacts that would contribute to the longer-term realization of increased participation of women in science and engineering. These assumptions were based on our experience, findings from other programs (e.g. other CWSEs, NSF ADVANCE) and the social science literature.[4]

By 2013, NSERC managers approached the CWSE Chairs in preparation for a meeting at the Canadian Engineering Education Association meeting in Montreal with the NSERC Chairs in Design Engineering. The Chairs were asked on both sides to develop a common set of metrics for their programs. The Design Chairs and the CWSE Chairs met to discuss and develop some common metrics at the meeting with Serge Villemure and Jennifer Houghton attending from NSERC. The CWSE Chairs went on to develop the set of metrics as shown in Fig 5.5. These were tangible numbers they could record and update throughout the years, and these served them well for reports to NSERC and the Coast-to-Coast updates published by the network every two years, from 2014 onward.

A few years later, NSERC managers again approached the Chairholders to enlarge or deepen the evaluation methods to capture more of the qualitative and long-term outcomes of their programs. The National Network then took on the task of developing a more detailed performance management strategy through the expert advice of an evaluation professional, Jane Whynot, as mentioned in Chap. 4. Jane spent months in 2017 and 2018 learning about each Chair's programs through meetings at conferences and visits and observations of Chair events in each region. Together, she and the Chairs developed a performance management strategy that would address the multiple dimensions of the programs, based on integrating gender into a theory of change. The tool identified spheres of influence, targeted areas of action and change, desired outcomes, and indicators for reporting progress toward the desired outcomes, both quantitative and qualitative, with suggested reporting intervals and reporting responsibilities, specifically who would report (the

[4] Davidson, V. and Holmes, A., *Joining WISE Conversations: Strategies and Successes from CWSE-ON 2003 – 2011*, 2012.

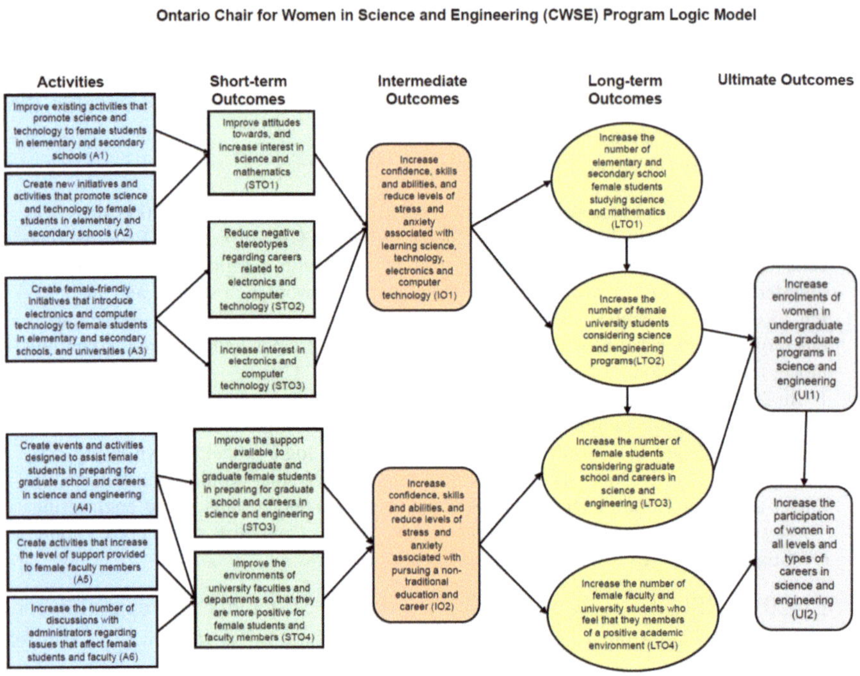

Fig 5.9 Valerie Davidson's Ontario Chair for Women in Science and Engineering Program Logic Model[1]

Chairs, the network or NSERC, or others beyond). For example, reporting on long-term (national) outcomes (over a decade or more) was expected to be the responsibility of others beyond the Chairholders, while they agreed to give input. The tool was delivered to NSERC as they were starting the first program evaluation for the CWSE program in 2020. Jane Whynot was hired as a consultant for the program evaluation exercise.

The methodology for the NSERC evaluation of the program is briefly described in the final report.[5] Not all Chairs could be interviewed in depth because of scope limitations on the effort. Six case studies were used, including interviews with four current regional Chairs, one early Chair and one representing the National Network. Partners were also interviewed bringing the total to 30 people. Eight other people, including program representatives, selection committee members, and some other Chairs were also interviewed as "key informants." A review of Chair applications and reports was conducted as well as a literature review and an environmental scan. Efforts were also hampered by the COVID-19 pandemic. The report admits that it focused on more recent activities and mentioned the lack of standardized methods among Chairholders to collect participant data. According to the report, "the

[5] NSERC, Evaluation of the Chairs for Women in Science and Engineering Program, 2021. https://www.nserc-crsng.gc.ca/_doc/Reports-Rapports/CWSE-Eval_e.pdf

evaluation also confirmed that measuring the overall impact of the CWSE program remains challenging." It found the new performance management strategy relevant and envisioned it "directly support[ing] future evaluations of the program."

Leveraging the Chair Program

The initial idea to have industry support for the national Chair proved to be most beneficial not only for leveraging any investment NSERC could put toward increasing the participation of women in science and engineering, but also as a mindset for seeking partners and broadening the scope of activities, be they outreach or research, through other grants. Already in the first generation, Chairholders reported additional funds raised through collaborative efforts such as Mary Williams' \$213,000 over the period 1997–2001, Claire Deschênes' \$1,100,210 over 1997–2006, and Elizabeth Cannon's \$425,000 through the "Imperial Oil Charitable Foundation, SSHRC, Alberta Innovation and Science and other government, professional and industry groups." Valerie Davidson reported \$712,150 in leveraged funds while the entire network reported \$7M for the period 2015–2020. Major SSHRC and NSERC CREATE grants led by the Chairholders or their close collaborators in the social sciences helped to boost these numbers. As major graduate training grants, the CREATE programs brought equity, diversity, and inclusion concepts, research, and activities to a broader swath of students and faculty members in the natural sciences, engineering, and medicine.

Close Connections

The Chairholders formed close connections not only as fellow pioneers in a unique program but also before becoming Chairs. As mentioned throughout the book, some future Chairholders were involved in previous Chair programs, such as Ève Langelier, who was an assistant Chair under Claire Deschênes' Chair from 2004 to 2006. Ève would wait ten years before becoming Chair herself in 2015. At UBC, the mentoring between Chairs was strong with Maria Klawe involving Anne Condon and Elizabeth Croft in her programs. Anne, being a computer scientist like Maria and in the same department, might have been a logical choice as a successor, but Elizabeth was in mechanical engineering. Her early involvement in Maria and Anne's Chairs gave her the motivation and experience to pursue the Chair position in the third round in the West. Elizabeth also involved Lesley Shannon in her activities which eventually led to her successful Chair bid in 2015. Mentoring between the generations also takes place across the country, with Mary Williams serving as a close ally to Valerie Davidson and being on the board of advisors for both Valerie and Catherine's Chairs in Ontario. Valerie had known Mary through a family connection while in her teens and benefited from her informal mentoring at key points in her education. At the ICWES conference in 2002, Mary encouraged Valerie to apply for the CWSE Chair and introduced her to Maria and Elizabeth. Valerie then served as a mentor to many of the third-generation Chairs while she continued to be involved with NSERC on matters related to the participation of women. Past Chairs could be counted on to help with support letters as they were probably the only people who knew the challenges of holding a CWSE Chair while trying to maintain an academic career in science and engineering. They gave generously of their time to help the next generations reach new heights (Fig. 5.10).

The visibility of the Chairs, namely through their titles and their proactive efforts such as running the ICWES 12 conference in 2002 in Ottawa, gave others the ideas for pursuing such a path. Catherine Mavriplis, for example, attended ICWES12 and later invited Maria Klawe to give the keynote talk in 2003 at her FORWARD to Professorship workshop in Washington, DC, and again in 2005 for the FORWARD workshop at MIT. When she eventually returned to Canada in 2008, Catherine had her eye on the Chair for Ontario. Similarly, Laleh Behjat was marked by meeting the Chairs at the 2013 Gender Summit in Washington and would later become Chair in 2020. In the competitive science and engineering academic world, taking on a Chair for Women is not a simple decision. All Chairholders would have had to think hard whether such a move would help or hinder their career. Having role models ahead of them, who were successful in their Chair efforts but also as science and engineering researchers, was enormously helpful in making that decision.

Once appointed as a Chair, the connections grew closer, especially with the help of the National Network funds which allowed for more frequent meetings of the Chairs and joint projects. Each generation bonded over a shared sense of

Fig. 5.10 The four West Coast Chairs at the closing ceremony for Elizabeth Croft's Chair in 2015. From left to right: Lesley Shannon (2015–2020), Elizabeth Croft (2010–2015), Anne Condon (2003–2010), and Maria Klawe (1997–2003)

responsibility and heavy load. The first generation of Chairs created a strong network advocating for policy change and initiating an international network through the creation of INWES. Their bonds remain strong as Monique and Claire have worked over 25 years together on various projects including the establishment of the Archive of Women in STEM, often with historian Ruby Heap. Almost all the past and current Chairs (13 out of 15) came together for the 2014 Archives workshop in Ottawa.

Through shared interests, Chairholders formed bonds, such as Annemieke and Catherine through hosting the Verna J. Kirkness Education Foundation program for Indigenous students, Annemieke and Tamara through their shared research on faculty climate across their regions, and more recently through the Network of Collaborations developed by Laleh and Shohini in which Catherine and Jennifer play a major role. The Network of Collaborations workshop at the Banff International Research Station in 2022 also included Annemieke, the independent Chair at Queen's University, Heidi-Lynn Ploeg, and ONWiE Chair Kim Jones. The Chairholders benefited from a steady and strong set of cheerleaders beyond the program such as Kim and others over the years. Chemistry professor Margaret-Ann Armour, University of Alberta, has been known as a vocal advocate for women in science since the early 1980s. She had founded WISEST (Women in Scholarship, Engineering, Science and Technology) and served as Associate Dean for Diversity in the Faculty of Science. She had also been President of the Women in Science, Engineering, Technology and the Trades (WinSETT) Centre, which she helped establish in 2010, involving the Chairs on the WinSETT Board and grant proposals to raise funds for WinSETT. WinSETT had a strong connection to WISE NL (Women in Science and Engineering Newfoundland) which had been close to Mary's Chair with the involvement of Carolyn Emerson and others. Gloria Montano from WISE NL and WinSETT and Nan Armour who had founded the Hypatia Project and been Chair of CCWESTT were ardent supporters from the Atlantic region. They interacted with the Chairs frequently through WinSETT and CCWESTT. Mary Wells also supported the Chairs through her various roles, starting with Associate Dean for Outreach at the University of Waterloo and ONWiE Chair, and then Dean of Engineering at Guelph University and now at Waterloo. Sadly, the Chairs lost two of their greatest mentors recently: Mary Williams passed away in 2015 and Margaret-Ann Armour passed away in 2019.

Reception in Canada and Abroad

While most Chairs felt somewhat unappreciated at their own institutions or units, mainly due to the stigma of doing "women's work" rather than "serious work" (with an undertone of lacking excellence, scholarship, and the hard work of disciplinary science or engineering research and the ensuing publications and grants), they definitely felt recognized by the greater community across Canada through invitations for keynote talks and partnering with like-minded activists from both similar and other (e.g., social science) disciplines. Several organizations routinely requested that the Chairs present their work, lead discussion groups, Chair conference sessions, and serve on their Boards of Directors (for example, WINSETT). The work was often divided between the five Chairs to handle the heavy load of requests. The National Network facilitated this sharing of responsibilities.

Several Chairs used their platform to also work internationally for the benefit of women in science and engineering. As early as 1995, Monique Frize traveled to Hong Kong, Kuala Lumpur, and Bangkok on a speaking tour organized by the University of British Columbia and the ASEAN Centre in Singapore to exchange information on how to increase the participation of women. Elizabeth Cannon presented papers in India and Greece as early as 2001. In 1999, Claire Deschênes took part in meetings organized by the United Nations Educational, Scientific and Cultural Organization (UNESCO) and the International Council for Science as a member of the Canadian delegation. With UNESCO's support, in 2002, Monique Frize, Claire Deschênes, and Gail Mattson from the United States founded the International Network of Women Engineers and Scientists (INWES). While the International Conference of Women Engineers and Scientists (ICWES) had existed since 1964, in 2002, INWES took over the management of the conference, and their first conference was held in Ottawa that year. All five first-generation Chairs were deeply involved and the Chair program was made visible on the international scene. Monique served as President of INWES from 2002 to 2008. By 2006, INWES counted more than 200,000 women as members. As of 2019, this number was 250,000 from 60 countries around the globe. Nadia Ghazzali currently serves as Deputy President of INWES (2021–2023). A new book entitled Women's Contribution to Science and Technology through ICWES conferences, authored by Monique Frize, Claire Deschênes, and Ruby Heap, is scheduled to be published by Springer this year (2024).

Claire Deschênes worked abroad, in Europe mostly, and through her time on the Canadian Commission for UNESCO and OECD, she noted that the European Union's "Ambassadors" program under the newly created European Platform of Women Scientists was inspired by the CWSE program. The second generation of Chairs also made a point of presenting their work internationally: Valerie Davidson traveled to France (to the Prometea Conference for Women in Engineering in 2009) and Australia (ICWES in 2011) to present papers on their behalf; Julita Vassileva did the same at a regional ICWES meeting in Poland in 2007. Nadia Ghazzali worked to set up INWES Arab World and l'Association des ingénieures et scientifiques marocaine (The Moroccan Association of Women Engineers and Scientists). She presented her work in Morocco often. Later, other Chairs attended international meetings in Europe, namely ICWES and Gender Equity in Higher Education conferences, to present the work of the Chairs. In 2009, Valerie Davidson co-organized the 2004 Women in Engineering Leadership Institute in Syracuse in the United States with Shobha Bhatia and Kelly Rusch, and Catherine Mavriplis regularly presented at US National Science Foundation (NSF) ADVANCE grantees meetings as Chair due to her previous involvement with the ADVANCE program as a grantee. The computer scientists among the Chairs played significant roles in the US-based

Association for Computing Machinery (ACM and ACM-W for women) and the Grace Hopper Celebration of Women in Computing Conference, with Anne Condon serving as program Chair for the very large conference in 2007.

In 2013, the NSERC CWSE program managers requested that the five Chairs have an important presence at the first Gender Summit in North America, in Washington, DC. There, the Chairs appeared as a panel discussing their activities and the program in general, as well as leading individual talks and workshops. The reception from summit participants was palpable. Many of the comments received from European participants reflected their awe at the breadth of activities undertaken by the Chairs and the economical investment by NSERC ($70,000/year/Chair) to produce all the work described. In 2017, NSERC CWSE program managers were heavily involved in the planning and organization of the second Gender Summit in North America, in Montreal. Nathalie Podeszfinski was the project manager for the summit. Ève Langelier, Chair for Quebec at the time, played a major role in the organization of the summit along with NSERC CWSE Program Manager Natalie Weiskopf. Again, the Chairs were asked to participate in full force: collectively they participated in three sessions and six posters.

A Training Ground for Leadership

Many of the CWSE Chairholders must either have been natural leaders and/or benefited from their tenure as Chairs to develop their leadership skills. They built trust and created networks that bought them credibility and got them noticed. Among the 16 Chairs profiled in the previous chapters, three have become university presidents, one has become provost, and several have become deans, associate deans, and associate vice presidents. Others have gone on to receive honorary doctorates and major scientific prizes.

Elizabeth Cannon, a first-generation Chair, became Dean of the School of Engineering at University Calgary in 2006 and President and Vice Chancellor in 2010, a role she kept until 2018. It is remarkable to think that Elizabeth was hired under the NSERC WFA (Women Faculty Awards) program (for women) and eventually made it to the level of university president. What would have happened without the WFA program? Elizabeth's career path is a major outcome for both the WFA and CWSE programs (Fig. 5.11).

Maria Klawe, also a first-generation Chair, became Dean of Engineering at Princeton University in New Jersey in the United States in 2003. In 2006, she took up the fifth presidency of Harvey Mudd College in Los Angeles in the United States, a position she held until July 2023. As president, she led the transformation of the College and, in particular, implemented changes to raise the number of women studying computer science from 10% to over 50% in less than ten years (Fig. 5.12).

Nadia Ghazzali, a second-generation Chair, was named President (Rectrice) of Université du Québec à Trois-Rivières in 2012 while she was still Chair. Negotiating the transition to a new institution and her new role, she managed to keep her Chair

Fig 5.11 Elizabeth Cannon addressing the University of Calgary as its eighth President

Fig. 5.12 Official photograph of the informal Maria Klawe as President of Harvey Mudd College

for two more years. Having been Vice President for Research in 2005–2006 and previously Vice Dean for Research at Université Laval, she had acquired significant leadership experience before taking up her CWSE Chair (Fig. 5.13).

Elizabeth Croft, a third-generation Chair, became Associate Dean, Education and Professional Development at UBC during her Chair tenure. In 2017, she became Senior Associate Dean at UBC and served as Dean of Engineering at Monash University in Melbourne in Australia from 2018 to 2022. Recently, she was appointed Vice President Academic and Provost at the University of Victoria in BC. Describing herself as someone who was once terrified of public speaking, she has built her leadership skills steadily through many challenging roles including the CWSE Chair (Fig. 5.14).

Claire Deschênes, a first-generation Chair, returned to continue and amplify her disciplinary research in fluid mechanics after her Chair ended, eventually winning the NSERC Synergy Prize in 2014 for her work in forming a consortium of hydraulic turbine companies to share data, test reduced scale models, and improve numerical modeling. She was then bestowed an honorary doctorate from the University of Ottawa in 2016, as was Monique in 1992, Elizabeth in 2013, and Maria in 2008. In

Fig. 5.13 Official photograph of Nadia Ghazzali at the time of her nomination as President of Université du Québec à Trois-Rivières in 2012

Fig. 5.14 The Twitter announcement by the University of Victoria President Kevin Hall on Elizabeth Croft's first day as Provost in 2022

2019, Université de Sherbrooke honored Claire with an honorary doctorate and the naming of its postdoctoral fellowship program, the Claire Deschênes Postdoctoral Fellowship Competition, aiming to ease the transition from graduate student to professorship for women in engineering and increasing the number of women engineering professors at their institution. The fellowship provides training for one year and a guaranteed faculty position at Université de Sherbrooke in engineering.

Other honorary doctorates were conferred to several of the Chairs: Monique received an honorary doctorate in Science from York University in 1994, from Lakehead University in 1995, and Mount Saint Vincent University in 2006; Elizabeth Cannon received honorary doctorates from Acadia University in 2010, Université de Montréal in 2016, and the University of Toronto in 2021; Maria Klawe received honorary doctorates from Ryerson University in 2001, the University of Waterloo in 2003, Queen's University in 2004, the University of Alberta in 2007, UBC in 2010, the University of Toronto in 2015, Concordia University in 2016, and McGill University in 2018; Valerie Davidson received an honorary doctorate from the University of Waterloo in 2022.

In 2001, Mary Williams, a first-generation Chair, had submitted a successful proposal for a second term as Chair from 2002 to 2005. In the interim, she was appointed Director General for the National Research Council Institute for Marine Dynamics. Dr. Williams left the Chair to take up the new appointment on September 4, 2002, when the second term of CWSE would have begun.

Annemieke Farenhorst, a third-generation Chair, became Associate Dean for Research in the Faculty of Agricultural and Food Sciences at the University of Manitoba toward the end of her Chair and was named Associate Vice President for Research in 2021.

The Changing Look of the Chair Programs

Logos for the Chair programs were not a must but proved effective in what we now call "branding" of activities: essentially becoming recognizable to the community. The logos below show an evolution in how the Chairholders perceived their roles and the participants they sought to engage. Lettering, color choices, and images of women all evolved with the times. Websites also became more sophisticated as the years went by. Sadly, many of them have been taken down: as the technology evolves, it is not feasible to maintain old sites (Fig. 5.15, 5.16, 5.17, 5.18, 5.19, 5.20, 5.21, 5.22, 5.23, 5.24, 5.25, 5.26, 5.27, 5.28, 5.29, and 5.30).

The National Network also developed its own logo, starting in the second generation in 2009 and updating it with the NSERC logo in the second version in 2015. The current logo reflects NSERC's new logo (Figs. 5.31, 5.32, and 5.33).

The National Network started in 2012 with a Coast-to-Coast Update brochure to be published every two years as an information resource to be distributed mainly at the biennial CCWESTT conferences. The first edition was text-heavy and had

Fig. 5.15 Mary Williams' Chair logo (1997–2022)

CWSE
Memorial University of Newfoundland
NSERC/PETRO-CANADA CHAIR
Women In Science and Engineering

Fig. 5.16 Claire Deschênes' Chair logo (1997–2006)

Fig. 5.17 Cecilia Moloney's Chair logo (2003–2010)

Fig. 5.18 Valerie Davidson's Chair logo (2003–2011)

Fig. 5.19 Julita Vassileva's Chair logo (2004–2011)

Fig. 5.20 Anne Condon's Chair logo (2003–2010)

Fig. 5.21 Nadia Ghazzali's Chair logo (2007–2014)

Fig. 5.22 Elizabeth Croft's Chair logo (2010–2015)

Fig. 5.23 Annemieke Farenhorst's Chair logo (2011–2020)

Fig. 5.24 Catherine Mavriplis's Chair logos: 2011–2016 (left) and 2017–2021 (right)

Fig. 5.25 Tamara Franz-Odendaal's Chair logo (2011–2023)

Fig. 5.26 Lesley Shannon's Chair logo (2015–2020)

Fig. 5.27 Ève Langelier's Chair logo (2015–)

Fig. 5.28 Laleh Behjat's Chair logo (2020–)

Fig. 5.29 Shohini Ghose's Chair logo (2020–)

Fig. 5.30 Jennifer Jakobi's Chair logo (2021–)

Fig. 5.31 National Network of CWSE Chairs logo (2009–2014)

Fig. 5.32 National Network of CWSE Chairs logo (2014–2019)

Fig. 5.33 National Network of CWSE Chairs logo (2020–)

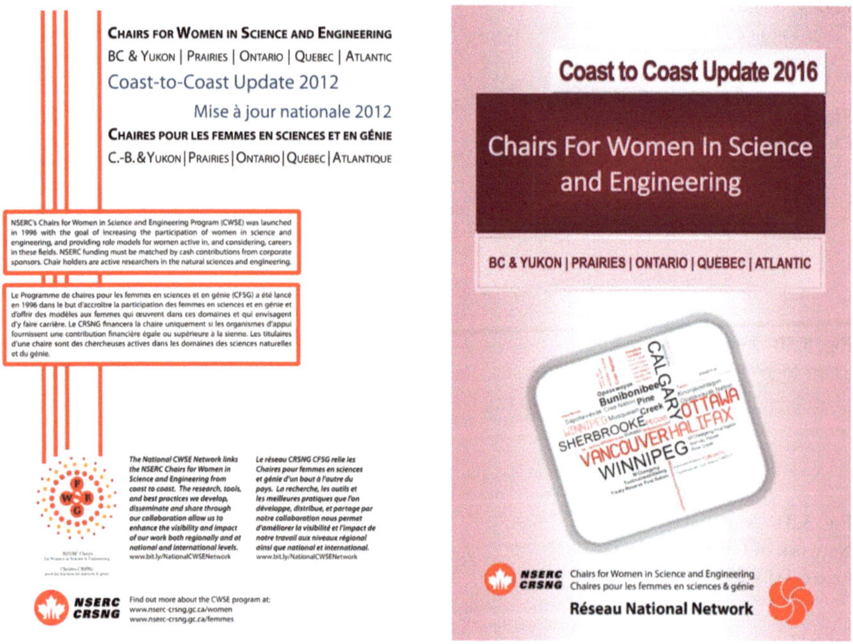

Fig. 5.34 Coast-to-Coast Update brochure cover designs: 2012 (left) and 2016 (right)

rudimentary graphic design due to budget constraints. The covers of the brochures below show a progression toward a more professional look (Fig. 5.34 and 5.35).

A long-standing problem for the Chairs was the hosting of a National Network website. In many conversations with NSERC program managers, the Chairholders sought to secure a link on the NSERC site for a common website. Finally, in 2018, a page for the Network was constructed under the Ontario Chair website, and in

Fig. 5.35 Coast-to-Coast Update brochure cover designs: 2017 (left) and 2018 (right)

2020, an independent website was created as https://stem-inclusion-stim.ca/ . It houses important resources, instructions for applying for the Dr. Margaret-Ann Armour National Conference Grant, and links to regional Chairs' websites.

Organizing the flow of information for the work of the Chairs was difficult due to the broad scope of the Chairs program. Most websites were organized in a linear fashion, evolving to a model which sought to engage the different target audiences, for example, youth, parents, teachers, university students, and professionals. The use of videos increased as the technology became more accessible. Tamara's Chair had partnered early on with Techsploration, a non-profit organization working with girls in Grades 9–12, who had started a powerful video series of women in STEM jobs. Tamara created 14 videos over the course of her Chair, and Catherine created nine, with three in French (Fig. 5.36).

The scope of the work of the Chairs was well captured by Valerie Davidson's depiction of her collaborators and the network she had built with her Chair. This and other graphic designs in her "Joining WISE Conversations" final report inspired the next generation of Chairs to amplify their messages and reach through bold and beautiful designs (Fig. 5.37). For example, Elizabeth Croft's infographic resource sheets were collected into a hardcover book entitled WWEST's Gender Diversity in STEM that was distributed across the country (Fig. 5.38).

Fig. 5.36 Video produced by Tamara's Chair in collaboration with ACENET and Techsploration about the role of an oceanographer, introducing Katja Fennel

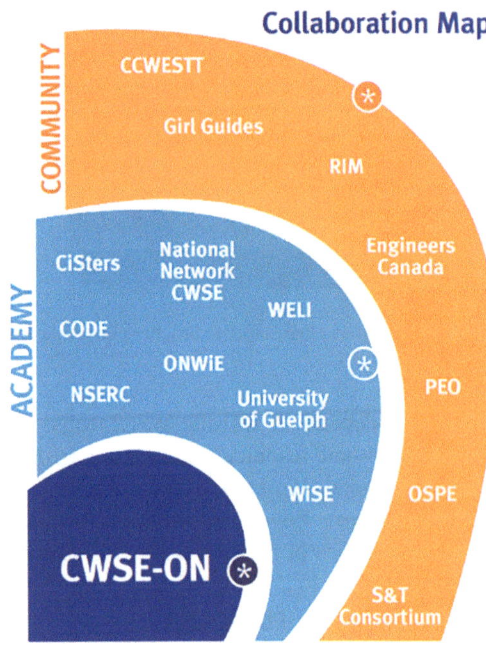

Fig. 5.37 Valerie Davidson's CWSE-ON program Collaboration Map

The Changing Landscape

As the first woman to earn a Bachelor's degree in Electrical Engineering at the University of Ottawa in 1966, Monique Frize had the courage to persevere in the unfamiliar territory of science and engineering despite being the only woman in most circumstances. In 1967, the Royal Commission on the Status of Women was just being formed with the first woman aeronautical engineer in the world, Elsie MacGill, playing a central part. By 1973, women made up only 3% of Bachelor of Engineering and/or Applied Science graduates. Following the pressure and

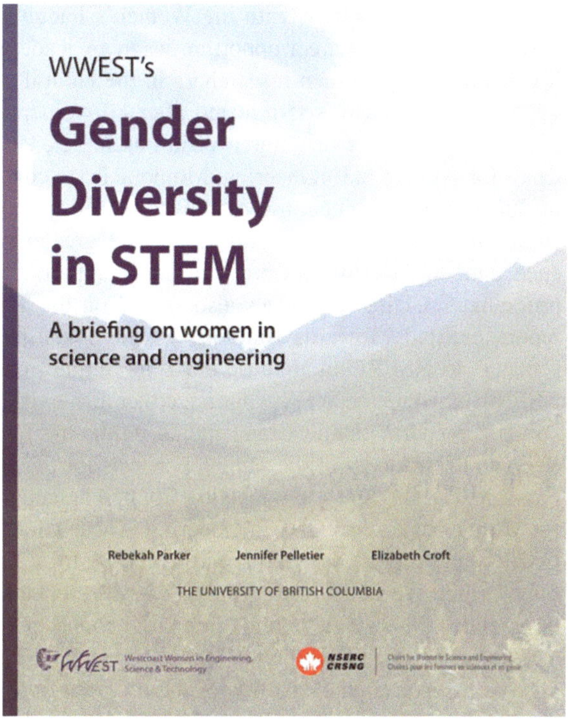

Fig. 5.38 Cover of Elizabeth Croft's Gender Diversity in STEM book

recommendations of the Royal Commission and the National Action Committee on the Status of Women, the 1970s and 1980s saw increases in the number of women studying and working in the sciences and engineering but the numbers remained very low: by 1989, when the first NSERC Chair for Women was established, only 13% of engineering graduates were women and only 2.9% of registered professional engineers were women. As we've seen, Claudette MacKay-Lassonde's leadership in establishing the first WISE (Women in Science and Engineering) association in 1977 and organizing the first Canadian Convention of Women Engineers in 1981 catalyzed a number of women into forming organizations to support women in these fields. As president of Professional Engineers Ontario, Claudette, along with Micheline Bouchard at the head of the Ordre des Ingénieurs du Québec, was able to engage the mainstream organizations of engineers throughout the 1980s, culminating in the first NSERC program aimed at increasing the participation of women and girls in engineering, and later the sciences too. Supported by these organizations, Monique Frize had a formidable task to address the entire country's needs to attend to the underrepresentation of women in science and engineering. The 1989 Polytechnique massacre, only months after her Chair was established, threw her efforts and the national focus on the state of women in technical fields into a brilliant limelight that no doubt contributed to the success of the launch of the NSERC Chair program.

Momentum was built with the Women's Faculty Award launched in 1991 by NSERC "to expand career opportunities in an academic setting for a selected number of promising women researchers in the natural sciences and engineering."[6] In 1995, the program was terminated after 80 women had received WFA awards to take up tenure-track or tenured positions in science or engineering. As National Chair for Women in Engineering, Monique Frize chaired the Task Force on Women in Science and Engineering, established "to advise Council on how to create an environment that facilitates and encourages greater participation of women in science and engineering research in Canada, particularly through NSERC's core programs."[7] Claire Deschênes also served on the Task Force which presented its report, entitled "Towards Building a New Scientific and Engineering Culture in Canada" to NSERC in June of 1996. Its number one recommendation was "that NSERC reinstate the WFA program, either as it existed or in a modified format that builds on partnership arrangements with the public and private sectors."[8] Recommendation 19 of the same report was for the creation of the regional CWSE program: "At the June 1996 meeting Council agreed to create up to four new Chairs for Women in Science and Engineering. These Chairs would be jointly funded by NSERC and the private sector, and would be for a five-year term, renewable once for a further five years. A program will be developed by NSERC staff over the coming months and a single competition will be held in the fall of 1996."

The Task Force recommended following up on WFA grant recipients and Council agreed to perform an evaluation of the program to find out if the grants were "real drivers for the hiring of women faculty." With the positive conclusions of the 1997 evaluation, the program was resumed as the University Faculty Awards (UFA) in 1999 and in 2001–2002, included the hiring of "Aboriginal people." On average, 21–26 UFA grants were awarded between 1999 and 2004, providing salary and research funds for the recipients. In 2003–2004, UFA expenditures were $3.8 million. By 2004, 113 women and two "Aboriginal" candidates had received a UFA award. Despite enthusiastic support from recipients and university administrators, a 2004 evaluation report found that the program was "not having a significant impact on the number of women hired for tenured or tenure-track positions or the number of women who choose to pursue a career in academia." Many thought the women recipients would have been hired anyway without the program, based on their high quality and success in the NSERC Discovery grant program. The major recommendation of the 2006 report was to discontinue the UFA program. NSERC management agreed and ran the final UFA competition in 2008 while committing to developing separate programs aimed at "women and Aboriginals."

---

[6] Ekos Research Associates, Evaluation of the University Faculty Awards Program, Final Report, June 2006.

[7] Thomas A. Brzustowski, President of NSERC, message upon release of the report of NSERC's Task Force on Women in Science and Engineering, June 12, 1996.

[8] NSERC Task Force on Women in Science and Engineering, Towards Building a New Scientific and Engineering Culture in Canada, 1996.

While the number of women faculty members rose through the UFA and WFA programs and the enrollments of women in university programs also grew, it was appropriate for the Chairs program to start multiple interventions through the five Chairs distributed regionally across the country in 1997. Definitely, the women who held the Chairs and the activists in the WISE and other scientific organizations devoted to recruiting and supporting women in these fields knew that the solutions to the underrepresentation problem were not just a numbers game. Interventions, designed to change the narrative about these fields, providing positive experiences for newcomers to the fields and providing information and critical support at junctions in the "leaky pipeline," came at an opportune time to sustain the women who had been recently hired or had recently graduated and grow the study programs. Designed by women who had faced similar situations, the NSERC Chair programs were a creative outlet for these role models in solving problems they not only perceived but had also lived. The relatively freeform mandate was key to their success and motivation to do this extra "women's work." Certainly, each of the Chairs went after their goals with passion and enthusiasm, only frustrated by the impossibly large responsibility they saw for themselves to help in all aspects of the problem.

From the start, the first generation of Chairs attacked the problem at all levels: girls in schools, parents, teachers, and through to university degree programs, professionals, and leadership positions. Early on, several Chairs concentrated some of their efforts on diverse populations of women, such as those in Indigenous communities, in particular in the Prairies. Now, the practice has broadened to almost every Chair. From the start, there was also a focus on conducting research on the status of women, generally in the form of statistics, but later as climate studies and eventually more specialized investigations in collaboration with social scientists and other researchers beyond the natural sciences and engineering. Leveraging the Chair funds and activities to apply for Social Sciences and Humanities Research Council (SSHRC) awards starting in the 2000s was an innovative yet necessary way to pursue the problem of underrepresentation of women and girls. Early on, the first generation of Chairs also saw the value of creating leadership and communications workshops to support women in these fields, such as the *Becoming Leaders* series and later conferences such as the 2000 New Frontiers, New Traditions Canadian Coalition for Women in Science, Engineering and Technology (CCWEST) conference. The Chairs have had a long-standing participation in the CCWEST conferences and have organized a significant number of them. These too have evolved, including issues of women in the trades since 2006, hence the updated CCWESTT designation.

In 2010, the Canada Research Excellence Chairs Program announced its first round of prestigious winners of multi-million-dollar grants. The incumbents were to be world-class researchers, and the program was designed to attract and retain high-caliber researchers to/in Canada. Among the 19 recipients, there were zero women. Industry Minister Tony Clement was vocal at the time about the situation. He requested that an ad hoc panel examine the selection process. The assembled Expert Panel on Women in University Research, led by Lorna Marsden, President Emerita of York University, included NSERC Chair for Women Nadia Ghazzali, at the time

President of Université du Québec à Trois-Rivières. Produced by the Council of Canadian Academies, the resulting 2012 report, Strengthening Canada's Research Capacity: the Gender Dimension,[9] provided a comprehensive collection of data, which was sorely lacking in Canada as remarked by the panel, focused on detailed measurements of women's progress at different ranks and in different disciplines rather than as aggregates, and identified "major policies and factors that affect the career trajectories of women researchers in Canada", noting that many occur before university or are embedded in institutional practices in our organizations and society. Most importantly, perhaps the key statement was that "rarely does one initiative work in isolation to 'fix' a problem of meet an objective. Rather it is a combination of practices and policies that, taken together, can be applied to meet objectives and bring about real change." The report was also reviewed by former NSERC Chairs, Valerie Davidson and Monique Frize.

In 2017, the Minister of Science, Kirsty Duncan, announced some changes to the Canada Research Chair program to enhance diversity in the program. Tier 1 Chairs were limited to a single renewal to a maximum of two seven-year terms, i.e., 14 years. Tier 2 Chairs could be converted to Tier 1 Chairs, and the distribution of Chair allocations was revised. Universities were also given a deadline of December 2017 to submit equity, diversity, and inclusion (EDI) plans to address equity targets. Things were heating up. By forcing universities to comply with their stated diversity targets, the 2019 competition resulted in 47% women, 22% visible minorities, 5% persons with disabilities, and 4% Indigenous people among the awardees. In these same years, Minister Duncan also fostered the development of the Tri-agency Dimensions Program, based on multiple policy consultations in which the Chairs took part, and based in part on the successful Athena Scientific Women's Academic Network (SWAN) program in the United Kingdom. Institutions are encouraged to adopt the Dimensions charter to commit to improving equity, diversity, and inclusion across the board within their campuses. A cohort of 17 institutions competed for a pilot project to develop a recognition program for their efforts in the EDI sphere. "The Dimensions program addresses barriers and discrimination faced by equity-deserving groups, including women, Indigenous Peoples, persons with disabilities, members of visible/racialized minorities, and members of LGBTQ2+ communities, in all disciplines and at postsecondary institutions – CEGEPS, colleges, polytechnics and universities," according to NSERC's website.

In 2019, the NSERC Chairs were involved in the first evaluation of the NSERC CWSE program. They were interviewed by the evaluators and wondered if the program, in light of a national, and to some extent international, migration away from women's issues and toward more encompassing EDI issues might mean the end of the CWSE program. The NSERC Chairs for Design Engineering program had recently been discontinued in 2019. The ensuing 2021 evaluation report of the CWSE program recommended that the program be continued as an important asset

[9] Council of Canadian Academies, Strengthening Canada's Research Capacity: The Gender Dimension / The Expert Panel on Women in University Research, 2012.

in NSERC's portfolio of initiatives to promote the participation of women in science and engineering. Shortly thereafter, however, NSERC CWSE program managers approached the current Chairs with new models for the Chair. While some of the current Chairs had been heavily involved in the Dimensions and other EDI programs, others had devoted significant energies to supporting Indigenous or Black communities, for example, Atlantic Chair Tamara Franz-Odendaal was instrumental in launching and supporting the Black Scientists Network. So, it was not a total surprise when NSERC announced that the next Atlantic Chair competition would be run as a pilot program for two years as the Chairs for Inclusion in Science and Engineering (CISE). The model changed to a team of three CISE Chairs at multiple universities across the Atlantic region, addressing long-standing concerns with the difficulty in covering several provinces as a single Chair. The program mentions multiple underrepresented groups and expects intersectionality to be addressed. At least one of the three applicants in a team must "self-identify" a woman. And at least one action must be focused on women. All-women teams are eligible.[10] The budget has been generously increased to $130,000 per year for two years with up to $70,000 in matching funds. The postdoctoral funds are still available with matching funds from the institutions, and an optional NSERC PromoScience supplement of $30,000 is also available for the team for "hands-on, interactive" activities for youth. At the time of writing this book, the first CISE Chairs are about to be announced. In the meantime, the National Network has updated their message with a new tagline: "Creating equitable systems to make meaningful change." Time will tell whether the program will evolve away from the focus on women, and if so, what effect that will have on the participation of women in STEM, or if there will be another course correction.

For further reading on the NSERC Chairs for Women in Science and Engineering, the reader is directed to:

- The Canadian Archive for Women in STEM at the University of Ottawa: https://www.uottawa.ca/library/archives-special-collections/canadian-archive-of-women-in-stem
- The Atrium repository at the University of Guelph: https://atrium.lib.uoguelph.ca/home

[10] NSERC, NSERC Chairs for Inclusion in Science and Engineering (pilot), 2023. https://www.nserc-crsng.gc.ca/Professors-Professeurs/CFS-PCP/CISE-CWSE_eng.asp

Index

C. D'Amours et al., *The Chairs for Women in Science and Engineering Program*, Women in Engineering and Science, https://doi.org/10.1007/978-3-031-54085-1

D

MIX
Papier aus verantwortungsvollen Quellen
Paper from responsible sources
FSC® C105338

If you have any concerns about our products,
you can contact us on
ProductSafety@springernature.com

In case Publisher is established outside the EU,
the EU authorized representative is:
Springer Nature Customer Service Center GmbH
Europaplatz 3, 69115 Heidelberg, Germany

Printed by Libri Plureos GmbH
in Hamburg, Germany